# ••••••• What's That Smell?

## Art Credits

Photograph of rescue team, FEMA News Photo by Andrea Booher, courtesy of the Federal Emergency Management Agency.

Photograph of George Aldrich courtesy of NASA.

Photograph of hunting items provided by Robinson Laboratories, Inc.

Elephant photograph courtesy of The Elephant Sanctuary.

Zits comic reprinted with special permission of King Features Syndicate.

Halimeter® photograph provided by Interscan Corporation.

Sniff port gas chromatograph/mass spectrometer photograph provided by The Procter & Gamble Company.

Photograph of roll top tub courtesy of Antique Bath Boutique.

Photograph of 4711 cologne provided by Cosmopolitan Cosmetics GmbH.

# What's That Smell?
## The Science Behind Adolescent Odors

**Editor**

**Mickey Sarquis,** Director
  Center for Chemistry Education

**Contributing Authors**

**Dianne Epp,** Lincoln East High School, Lincoln, NE

**Susan Hershberger,** Department of Chemistry and Biochemistry
  Miami University, Oxford, OH

**Jerry Sarquis,** Department of Chemistry and Biochemistry
  Miami University, Oxford, OH

**Virginia Wysong,** Tri-County North Middle School, Lewisburg, OH

**Terrific Science Press**
**Miami University Middletown**
**Middletown, Ohio**

Terrific Science Press
Miami University Middletown
4200 East University Boulevard
Middletown, OH 45042
513/727-3269
cce@muohio.edu
www.terrificscience.org

© 2003 by Terrific Science Press™
All rights reserved
Printed in the United States of America

10  9  8  7  6  5  4  3  2  1

This monograph is intended for use by teachers, chemists, and properly supervised students. Users must follow procedures for the safe handling, use, and disposal of chemicals in accordance with local, state, federal, and institutional requirements. The cautions, warnings, and safety reminders associated with experiments and activities involving the use of chemicals and equipment contained in this publication have been compiled from sources believed to be reliable and to represent the best opinions on the subject as of the date of publication. Federal, state, local, or institutional standards, codes, and regulations should be followed and supersede information found in this monograph or its references. The user should check existing regulations as they are updated. No warranty, guarantee, or representation is made by the authors or by Terrific Science Press as to the correctness or sufficiency of any information herein. Neither the authors nor the publisher assume any responsibility or liability for the use of the information herein, nor can it be assumed that all necessary warnings and precautionary measures are contained in this publication. Other or additional information or measures may be required or desirable because of particular or exceptional conditions or circumstances or because of new or changed legislation.

ISBN: 1-883822-27-0

The publisher takes no responsibility for the use of any materials or methods described in this monograph, nor for the products thereof. Permission is granted to copy the materials for classroom use.

This material is based upon work supported by the **National Science Foundation** (Grant Numbers 9950011, 9355523, and 9153930); **National Institute of Environmental Health Sciences** (Grant Number IR 25 ESO8192-01); and the **Ohio Board of Regents** (Grant Number 01-37). Any opinions, findings, and conclusions or recommendations expressed in this material are those of the authors and do not necessarily reflect the views of the funding agencies.

# Table of Contents

Acknowledgments ..................................................................................................... vii
Getting the Most from this Monograph ................................................................... viii

## Chapter 1: Teacher Background ........................................................... 1
The Sense of Smell ...................................................................................................... 2
Chemicals with Odors ................................................................................................. 6
Applications of Smell and Odors ............................................................................... 10
Body Scents ................................................................................................................ 14

## Chapter 2: The Nose Knows—Exploring the Sense of Smell ............ 17
Chapter 2 Overview ................................................................................................... 18
National Science Education Standards .................................................................... 18
Cross-Curricular Integration ..................................................................................... 19
Section 1 Background: How Your Sense of Smell Works ........................................ 21
Section 1 Experiment: Identifying Odors ................................................................. 23
Instructor Notes for Section 1 .................................................................................. 25
Section 2 Background: Knowing Your Sense of Smell ............................................ 30
Section 2 Experiment: Get a Whiff of This ............................................................... 32
Instructor Notes for Section 2 .................................................................................. 34
Section 3 Background: Molecules with a Message ................................................. 36
Section 3 Experiment: Molecules You Can Smell .................................................... 38
Instructor Notes for Section 3 .................................................................................. 40

## Chapter 3: The Origins of Body Odor ................................................. 45
Chapter 3 Overview ................................................................................................... 46
National Science Education Standards .................................................................... 46
Cross-Curricular Integration ..................................................................................... 48
Section 1 Background: Your Unique Scent .............................................................. 49
Section 1 Experiment: Scent Detectors .................................................................... 51
Instructor Notes for Section 1 .................................................................................. 53
Section 2 Background: Everything You Always Wanted to Know About Sweat ... 56
Section 2 Experiment: Using a Sweat Meter ............................................................ 58
Instructor Notes for Section 2 .................................................................................. 59

Section 3 Background: No Sweat ........................................................................... 62
Section 3 Experiment: Sweating Is Cool .............................................................. 63
Instructor Notes for Section 3 ............................................................................... 65

## Chapter 4: When Life Stinks ............................................................... 71
Chapter 4 Overview .................................................................................................. 72
National Science Education Standards ............................................................... 72
Cross-Curricular Integration .................................................................................. 73
Section 1 Background: Sweaty Feet .................................................................... 75
Section 1 Experiment: Ban the Rotten Sneaker ................................................. 77
Instructor Notes for Section 1 ............................................................................... 80
Section 2 Background: Beware the Fire-Breathing Dragon! ............................ 84
Section 2 Experiment: Spray It Away .................................................................... 86
Instructor Notes for Section 2 ............................................................................... 89
Section 3 Background: Control and Conquer .................................................... 92
Section 3 Experiment: Life in the Pits .................................................................. 94
Instructor Notes for Section 3 ............................................................................... 96

## Chapter 5: Combating and Controlling Body Odor .......................... 99
Chapter 5 Overview ................................................................................................ 100
National Science Education Standards ............................................................. 100
Cross-Curricular Integration ................................................................................ 101
Section 1 Background A: The History of Bathing ............................................ 102
Section 1 Background B: Perfumes and Sensitivity ........................................ 104
Section 1 Experiment: Pick a Fragrance ............................................................ 106
Instructor Notes for Section 1 ............................................................................. 108
Section 2 Background: Scent Masquerade ....................................................... 111
Section 2 Experiment: The All-American Cover-Up ........................................ 113
Instructor Notes for Section 2 ............................................................................. 115

## Bibliography ......................................................................................... 119

# ● Acknowledgments

The authors and editor wish to thank the following individuals who have contributed to the development of *What's That Smell? The Science Behind Adolescent Odors*.

## Terrific Science Press Design and Production Team

*Document Production Managers:* Amy Stander and Susan Gertz
*Production Coordinator:* Dot Lyon
*Technical Writing and Editing:* Dot Lyon and Tom Schaffner
*Copy Editing:* Kate McCann, Jeri Moore, Becky Franklin
*Production:* Tom Schaffner, Dot Lyon, Jeri Moore, Kate McCann, Dawnetta Chapman, Barbara Egan
*Illustrations:* Carole Katz and Susan Gertz
*Cover Design and Layout:* Susan Gertz
*Technical Review:* Lynn Hogue

## Content Specialists, Reviewers, and Classroom Testers

George Aldrich, NASA Johnson Space Center, White Sands Test Facility, Las Cruces, NM

Marina Canepa, Department of Chemistry and Biochemistry, Miami University Middletown, OH

Beverly Cowart, Monell Chemical Senses Center, Philadelphia, PA

Joyce Feltz, Center for Chemistry Education, Miami University Middletown, OH

Patrick Greco, Department of Chemistry, Sinclair Community College, Dayton, OH

Mary Beth Hogan, Cincinnati Hills Christian Academy, Cincinnati, OH

James Janik, Department of Zoology, Miami University Middletown, OH

Kevin Kittredge, Department of Chemistry and Biochemistry, Miami University Middletown, OH

John Loper, Department of Molecular Genetics, Biochemistry, and Microbiology (active emeritus), University of Cincinnati, OH

J. Timothy Perry, Mt. Hebron High School, Ellicott City, MD

Ravi Ranatunga, The Procter & Gamble Company, Cincinnati, OH

George Rizzi, The Procter & Gamble Company (retired), Cincinnati, OH

Heather Rocchetta, The Procter & Gamble Company, Cincinnati, OH

Sandy Van Natta, White Oak Middle School, White Oak, OH

Jon Witt, The Procter & Gamble Company, Cincinnati, OH

Linda Woodward, Affiliate for the Center for Chemistry Education, Miami University Middletown, OH

Dustin Yontz, Department of Chemistry and Biochemistry, Miami University, Oxford, OH

# Getting the Most from this Monograph

This section contains important information that will help you use this monograph effectively. The topics covered here include
- the purpose of this monograph and how it is organized;
- tips for doing the experiments; and
- important safety information.

## Purpose of this Monograph

This monograph serves as a teacher resource for middle and high school. The lessons are appropriate for science, health, and interdisciplinary classes. The purpose of the monograph is to help young teens understand some of the chemistry and biology of natural body odors. Eleven hands-on experiments help teens examine cultural choices for dealing with body odor and the science behind these choices, so that they can make informed decisions in their own personal hygiene regimens.

Teenagers undergoing puberty deal with many different changes in their lives. Their bodies are rapidly changing due to the normal processes of hormonal development; social and emotional factors complicate their lives even more. All of these factors, combined with the stresses and joys of everyday life, influence how teens feel about themselves. Therefore, a teen's physical changes, including his or her new body odors associated with hormonal changes, need to be dealt with in a culturally accepted manner. Keep in mind that this monograph presents personal hygiene from an American viewpoint; many other countries and cultures view body odors and hygiene very differently.

## How this Monograph Is Organized

Chapter 1 provides teachers with a strong foundation in the topics of smell and odors. The remaining four chapters are designed for classroom use and include backgrounds for teachers and students, reproducible student pages for hands-on student experiments, and instructor notes. Instructor notes provide materials lists, procedure notes, sample data, and answers to student questions. The CD-ROM included with this monograph contains electronic files of the student pages so that you can easily tailor them to meet your needs.

**Chapter 1,** "Teacher Background," contains information on the following topics:
- the sense of smell,
- chemicals with odors,
- applications of smell and odors, and
- body scents.

**Chapter 2,** "The Nose Knows—Exploring the Sense of Smell," contains the following experiments:
- *Section 1:* "Identifying Odors"—Students attempt to identify several common substances by odors alone.
- *Section 2:* "Get a Whiff of This"—Students take the alcohol sniff test to find out how keen their sense of smell is.
- *Section 3:* "Molecules You Can Smell"—Students identify unknown substances by odor and guess the familiar scent produced when the unknown odors are blended together.

**Chapter 3,** "The Origins of Body Odor," contains the following experiments:
- *Section 1:* "Scent Detectors"—Students attempt to guess the gender and identity of volunteers solely on the basis of hand odor.
- *Section 2:* "Using a Sweat Meter"—Students investigate factors that cause the cellophane fish to behave as it does.
- *Section 3:* "Sweating Is Cool"—Students learn how sweat cools the body by measuring temperature changes during the evaporation of different liquids.

**Chapter 4,** "When Life Stinks," contains the following experiments:
- *Section 1:* "Ban the Rotten Sneaker"—Students investigate how foot odor can be reduced by using shoe inserts.
- *Section 2:* "Spray It Away"—Students study the effectiveness of breath spray in covering odors.
- *Section 3:* "Life in the Pits"—Students simulate underarm conditions and explore how deodorants and antiperspirants work.

**Chapter 5,** "Combating and Controlling Body Odor," contains the following experiments:
- *Section 1:* "Pick a Fragrance"—Students compare classroom and gender preferences for a variety of perfumes, colognes, and aftershave lotions.
- *Section 2:* "The All-American Cover-Up"—Students explore the masking of odorous test substances and design an experiment to investigate the ability of antiperspirants to control moisture.

## Safety First

Experiments, demonstrations, and hands-on activities add relevance, fun, and excitement to science education at any level. However, even the simplest experiment can become dangerous when the proper safety precautions are ignored or when the experiment is done incorrectly or performed by students without proper supervision. While the experiments in this monograph include cautions, warnings, and safety reminders from sources believed to be reliable, and while the text has been extensively reviewed, it is your responsibility to develop and follow procedures for the safe execution of any experiment you choose to do. You are also responsible for the safe handling, use, and disposal of chemicals in accordance with local and state regulations and requirements.

- Read each experiment carefully and observe all safety precautions and disposal procedures. Determine and follow all local and state regulations and requirements.

- Always practice experiments yourself before using them with your class. This is the only way to become thoroughly familiar with an experiment, and familiarity will help prevent potentially hazardous (or merely embarrassing) mishaps. In addition, you may find variations that will make the experiment more meaningful to your students.

- You, your assistants, and any students participating in the preparation or performance of an experiment must wear appropriate personal protective equipment in the laboratory.

- Special safety instructions are not given for everyday classroom materials being used in a typical manner. Use common sense when working with hot, sharp, or breakable objects. Keep tables or desks covered to avoid stains. Keep spills cleaned up to avoid falls.

- Some students may have sensitivities to the fragrances used in some of the experiments. Such sensitivities can cause allergic reactions. Before conducting the experiments, find out if any students have fragrance sensitivities and take appropriate actions.

- Remember that you are a role model for your students—your attention to safety will help them develop good safety habits while assuring that everyone has fun with these experiments.

# Chapter 1
# Teacher Background

The Sense of Smell ............................................................................................2
    Seven Primary Odors ...............................................................................4
    Nasal Dysfunction....................................................................................5
    Perception of Smell in Western Culture ..................................................6
Chemicals with Odors........................................................................................6
Applications of Smell and Odors .................................................................... 10
    Using Smell to Detect Illness................................................................. 10
    Aromatherapy........................................................................................ 10
    Fragrance Industry ................................................................................ 11
    Electronic Noses ................................................................................... 12
    Development of the Stink Bomb ........................................................... 13
Body Scents..................................................................................................... 14

# The Sense of Smell

Of the five senses, smell is the least understood. The location of the olfactory membrane high within the nose makes studying the sense of smell difficult. In addition, smell is a subjective phenomenon that is difficult to measure. Smell is also less developed in human beings than in typical laboratory animals, reducing the usefulness of animal models in investigations of smell.

The olfactory membrane is a layer of cells on the roof of the nasal cavity. (See Figure 1-1.) The membrane consists of three types of cells: supporting cells, basal cells, and olfactory neurons. Sources vary widely on the number of olfactory neurons that humans have; some indicate that humans have 5–50 million olfactory neurons.

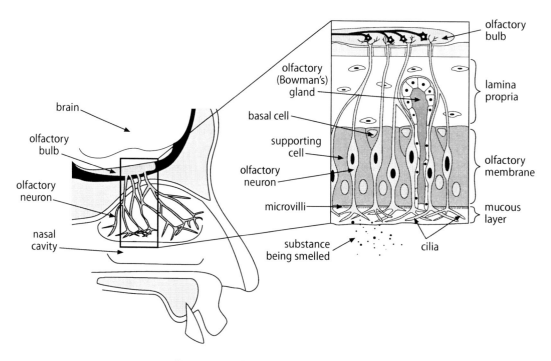

*Figure 1-1: Cross-section of the face*

An olfactory neuron is a bipolar neuron that consists of a cell body with two extensions: a dendrite that serves to input information and an axon that carries information toward the brain. (See Figure 1-2.) The axon of each olfactory neuron feeds into an olfactory bulb at the base of the brain. The dendrite of the neuron branches out into about eight hairlike cilia that extend into the nasal cavity. Secretions from olfactory (Bowman's) glands form a layer of moist mucus around the cilia.

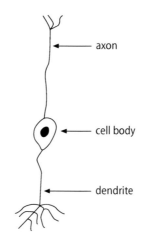

*Figure 1-2: General structure of a bipolar neuron*

The process of smelling begins when airborne molecules breathed into the nose are dissolved in the mucous layer surrounding the olfactory cilia. The odorant molecules interact with specific olfactory-receptor proteins that span the neuron cell membranes. The olfactory neurons react to these stimuli and in the process, initiate nerve impulses. The impulses pass through the neurons to one of the olfactory bulbs at the base of the brain. The olfactory bulb sends signals to the olfactory cortex, the hypothalamus, and portions of the limbic system. These areas of the brain interpret the signal to complete the smelling process.

In general, neurons that are lost from the nervous system are gone forever. Olfactory neurons are a major exception. Olfactory basal cells give rise to replacement olfactory neurons about every 30–60 days. As a result, neurons damaged by, for example, inhalation of strong irritants are replaced and the sense of smell can recover.

The nose may contain 400 or more different kinds of olfactory-receptor proteins that together can recognize about 10,000 distinct odors. With fewer kinds of receptors than distinct recognizable odors, how does the nose distinguish between all those different odors? Results from several laboratories indicate that each olfactory receptor can bind to a different subset of odorant molecules at different concentrations and that each odor binds to several different receptors. Thus, our noses use a combinatorial receptor coding scheme, meaning that different odors activate different combinations of odor receptors.

Genomic sequencing has resulted in more detailed studies of the olfactory-receptor genes. The human genome has about 1,000 sequence-related genes, a gene family of olfactory-receptor genes located in several clusters in the chromosomes. Of these, about 400 appear to code for functional olfactory-receptor proteins, while the other several hundred sequences are pseudogenes that do not code for complete olfactory receptors. Humans show individual variations in the sequence and in the number of copies of a particular olfactory-receptor gene, which may help account for differences in individuals' sense of smell.

In contrast to humans, the mouse olfactory-receptor gene family contains about 1,500 sequences, and a lower percentage of these appear to be pseudogenes. As a result, the pool of functional olfactory-receptor genes in the mouse appears to be more than three times larger than that in humans. Sequencing of the dog genome is currently in progress. Comparisons of the quantities of dog and human functional genes should prove interesting since the dog's sense of smell has been estimated to be anywhere from 1,000–10,000 times better than the human's sense of smell.

After constant exposure to the same odor, olfactory fatigue occurs and the odor stimulus fails to continue to trigger the flow of electrical signals to the brain. Some reports indicate that olfactory fatigue can occur within 60 seconds of smelling a particular odor. This fatigue can cause a dangerous situation when a person becomes accustomed to the odor of a toxic chemical. For example, a person exposed to high concentrations of hydrogen sulfide, a poisonous gas with a characteristic rotten egg odor, can experience olfactory fatigue and not continue to be warned of toxic levels in the air. Such fatigue is not limited to the sense of smell. For example, the nerves in our skin do not detect the constant contact of our clothing.

## Seven Primary Odors

Sensory research has resulted in the classification of sensory organ sensations. For example, the tongue detects four primary taste sensations (sour, salty, sweet, and bitter) and the eye detects three primary color sensations (blue, green, and red). However, classifying the primary odors detected by the nose has been much more difficult. Many systems for classifying odors have been proposed. The well-known classification of odors listed in the following table has been attributed to both Carolus Linnaeus (1707–1778) and John Amoore (1930–1998). Many systems also include cheesy and buttery classifications.

| Primary Odors | |
|---|---|
| Odor | Example |
| camphoraceous | Noxema® skin cream |
| musky | musk oil |
| floral | roses |
| pepperminty | mint candy |
| ethereal | dry-cleaning fluid |
| pungent | vinegar |
| putrid | rotten eggs |

## Nasal Dysfunction

Not all noses work equally well. More than two million Americans (about 1 percent of the population) suffer from a loss of the sense of smell. This condition is called anosmia and can be caused by many disorders, including inflammatory nasal disease, upper respiratory tract infection, aging, genetic defect, viral infection, allergy, nasal polyps, head injury, mental disorder, and dementia. Anosmia can also be a side effect of prescription drugs. Some of these conditions can be successfully treated with antibiotics, decongestants, antihistamines, steroids, other nasal drugs, trace metals, or surgery. Anosmia can result when olfactory neurons are damaged due to head trauma. Anosmia caused by head injuries may be temporary or permanent.

Other disorders of the nose include parosmia and phantosmia. People with parosmia have a distorted sense of smell. For example, they may smell a banana and say that it smells like rotting flesh. By comparison, people with phantosmia experience olfactory hallucinations. These people sometimes perceive an odor (usually unpleasant) when no odorant is present.

Contrary to popular belief, research shows that blind people do not actually have a keener sense of smell than the rest of the population. Research also shows that the blind, as well as others, can be trained to enhance their smelling performance. Being exposed repeatedly to certain smells may activate receptor cells and improve smelling. However, for such training to work, people must have the genes to produce the necessary receptors and the receptors must be in working order. Therefore, smelling disorders cannot be cured by this training method.

> **Class Research Idea**
> As a cross-curricular assignment, you may want to have students research cultural perceptions of smell and present their findings to the class.

## Perception of Smell in Western Culture

Some cultural historians have determined that, in the Western world, smell has a lower status than vision and hearing. The low status of smell is evident by our lack of olfactory terminology. Although we can describe something we see in the most minute detail, it is virtually impossible to describe an odor to someone who hasn't smelled it. While the average human can recognize about 10,000 different odors, we resort to crude analogies when asked to describe odors. For example, when we describe an odor as smoky, sulfurous, floral, or fruity, we are actually describing the odor in terms of something else. We also tend to describe an odor based on how it makes us feel (for example, disgusting, sickening, pleasurable, or delightful).

Many nonwestern cultures have placed a higher level of importance on odors and smell. In some Arab countries, breathing on a person you are speaking with signals friendship and goodwill. The Onge of the Andaman Islands in the Bay of Bengal define the universe and everything in it by odor. Their calender is based on the odors of flowers in bloom, and their seasons are named after specific odors.

Olfactory research is now attracting a large number of anthropologists, sociologists, and historians. This and other indicators may show that the Western world is raising the status of smell.

Remember that an odor is simply a perception triggered through a series of chemical reactions. The same chemical may be perceived very differently from one species to another or even from one human to another. A molecule that "smells bad" to a human may hold great appeal for a dung beetle.

# ● Chemicals with Odors

Chemicals that produce the sensation of odor share several common characteristics. Generally, odorous chemicals
- are soluble, at least marginally, in water and fat;
- are polar molecules (having a slightly positive region and a slightly negative region); and
- have low to intermediate molecular weights, typically between 15 and 300 amu (atomic mass units).

Most odorous chemicals are organic compounds, which means they are composed mainly of carbon and hydrogen atoms. The chemical bonds between the carbon and hydrogen atoms inside these molecules are strong. However, the forces holding one molecule close to another are rather weak. The weaker intermolecular attractions make these organic molecules rather easily vaporized so that their scents are detectable by the nose.

Organic compounds typically also include functional groups that contain oxygen, sulfur, or nitrogen. Certain functional groups are often associated with particular odors. (See the table on the following page for a list of some common examples.)

Inorganic compounds do not typically contain both carbon and hydrogen. They usually produce little or no odor because, unlike organic compounds, they are not easily vaporized. For example, in table salt, positively charged sodium ions ($Na^+$) are strongly attracted to negatively charged chloride ions ($Cl^-$); each $Na^+$ is surrounded by 6 $Cl^-$, and likewise, each $Cl^-$ is surrounded by 6 $Na^+$. These electrostatic ionic bonds are much stronger than the intermoleculer forces that hold organic molecules together.

A few inorganic compounds do have strong odors, including ammonia ($NH_3$), hydrogen sulfide ($H_2S$), and hydrochloric acid (HCl). Also, the family of elements known as halogens have characteristically pungent odors. These include fluorine ($F_2$), chlorine ($Cl_2$), bromine ($Br_2$), and iodine ($I_2$).

## Functional Groups and Structures of Some Odorous Chemicals

| Common Example | Odor | Structural Formula with Functional Group Shaded | Functional Group |
|---|---|---|---|
| isopropyl alcohol | antiseptic | $CH_3-\underset{\underset{CH_3}{\|}}{\overset{\overset{H}{\|}}{C}}-OH$ | alcohol |
| diethyl ether | pungent, sweetish | $CH_3-\underset{\underset{H}{\|}}{\overset{\overset{H}{\|}}{C}}-O-\underset{\underset{H}{\|}}{\overset{\overset{H}{\|}}{C}}-CH_3$ | ether |
| acetaldehyde | pungent, suffocating | $CH_3-\overset{\overset{O}{\|\|}}{C}-H$ | aldehyde |
| acetone | nail-polish remover | $H-\underset{\underset{H}{\|}}{\overset{\overset{H}{\|}}{C}}-\overset{\overset{O}{\|\|}}{C}-\underset{\underset{H}{\|}}{\overset{\overset{H}{\|}}{C}}-H$ | ketone |
| acetic acid | vinegar | $CH_3-\overset{\overset{O}{\|\|}}{C}-OH$ | carboxylic acid |
| octyl acetate | orange | $CH_3-\overset{\overset{O}{\|\|}}{C}-O-\underset{\underset{H}{\|}}{\overset{\overset{H}{\|}}{C}}-(CH_2)_6-CH_3$ | ester |
| trimethylamine | fishy | $H-\underset{\underset{H}{\|}}{\overset{\overset{H}{\|}}{C}}-\underset{\underset{H-\underset{\underset{H}{\|}}{\overset{\overset{}{\|}}{C}}-H}{\|}}{N}-\overset{\overset{H}{\|}}{\underset{\underset{H}{\|}}{C}}-H$ | amine |
| acetanilide | musty | phenyl$-\underset{\underset{H}{\|}}{N}-\overset{\overset{O}{\|\|}}{C}-CH_3$ | amide |
| methyl mercaptan | skunk | $H-\underset{\underset{H}{\|}}{\overset{\overset{H}{\|}}{C}}-SH$ | thiol or mercaptan |

Small differences in the geometric structure of molecules can mean dramatic differences in smell. For example, benzaldehyde has an aldehyde functional group and a strong odor of bitter almonds. By comparison, benzoic acid has a carboxylic acid functional group and is odorless. (See Figure 1-3.) The second example in Figure 1-3 shows a minor molecular difference between vanillin and ethyl vanillin. Due to this minor difference, the vanilla odor is three times stronger in ethyl vanillin than in vanillin. (Figure 1-3 uses common shorthand notation for the carbons and hydrogens of the benzene rings. Refer to Figure 2-4 in Chapter 2 for an explanation of the shorthand.)

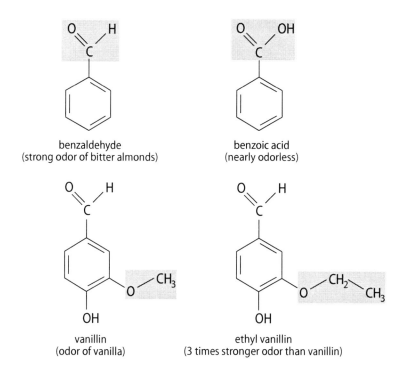

Figure 1-3: Examples of minor molecular differences
resulting in major odor differences
(Shading indicates the locations where molecular differences occur.)

Attractions between neighboring molecules also affect volatility and resulting odor strength. Benzoic acid is a prime example of this. Due to an especially strong intermolecular attraction, called hydrogen bonding, benzoic acid actually exists as a dimer (two molecules in close association with each other). (See Figure 1-4.) This dimer is far less volatile than a single molecule of benzaldehyde.

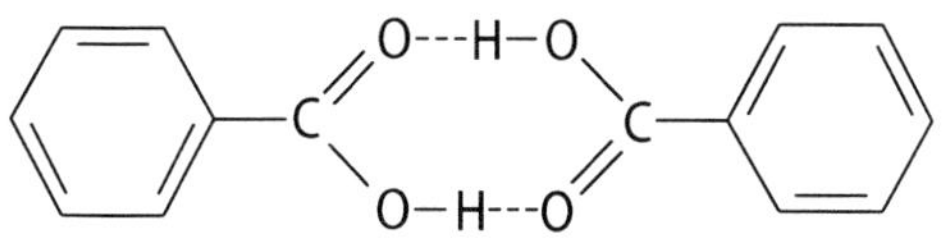

Figure 1-4: Benzoic acid dimer
(--- indicates hydrogen bonding.)

# ● Applications of Smell and Odors

Some clever real estate agents advise hopeful sellers to bake bread or simmer cinnamon sticks on the stove just before the house is shown to a prospective buyer because many people identify these smells with the comforts of home. This is an example of a potentially effective application of the sense of smell. Physicians, holistic medicine practitioners, and researchers for industry and government all use applications of the sense of smell to achieve widely diverse outcomes.

## Using Smell to Detect Illness

The observation that certain body odors can indicate illness dates back many centuries. For example, in about 400 BC the ancient Greek physician Hippocrates recommended sniffing a patient to provide a diagnosis. Many centuries later, the 11th century Arabian alchemist and physician Avicenna diagnosed patients by noting changes in the smell of their urine.

> **Class Research Idea**
> Have students research related topics about smell and odor, either on the Internet or at the library. Topics can include the ones found in this Teacher Background or others of interest to the class. Ask students to write a few paragraphs about the topic they researched. Have students present their findings to the class.

The observation that body odors indicate the presence of disease led many physicians to believe that these odors were the cause of the disease. Hippocrates burned scented stakes to combat the plague, and later, many physicians based their healing methods on the use of fragrance. Even through the early 19th century, fragrance was widely used to treat physical and mental disorders. These practices subsided during the 19th century and were replaced with the use of extracts, potions, and other medicines.

Today, some researchers and medical professionals still recognize that smell is an important tool in diagnosing disease. For example, breath odors can indicate gastrointestinal problems, diabetes, infections, sinus ailments, and liver problems. Odors from blood and urine can indicate liver and bladder problems. Some professionals say that people with diabetes smell like ketones, with the plague smell like mellow apples, with measles smell like freshly plucked feathers, with yellow fever smell like a butcher shop, and with nephritis smell like ammonia.

## Aromatherapy

Aromatherapy involves the use of essential oils (the essences from aromatic plants) for psychological and physical wellness. Many historians believe that the Egyptians were the first to use aromatherapy for therapeutic purposes. Ancient Egyptians massaged fragrant oils into the skin after bathing and used myrrh, lotus, and sandalwood oils in purification and embalming rituals. After

Greek physicians traveled to Egypt and learned about the oils, Greeks used aromatic oils for medicinal and cosmetic purposes. The Romans learned most of their medical knowledge from the Greeks and improved upon the concepts. They diffused essential oils in their temples and political buildings and soaked in oil-scented baths. They also imported new aromatic products from East India and Arabia.

The modern rediscovery of the healing power of essential oils is attributed to the French cosmetic chemist René-Maurice Gattefossé. After burning his arm in a laboratory explosion in 1910, Gattefossé dipped his arm in a vessel of lavender oil that he thought contained water. The burn began healing quickly, and regular applications of lavender oil seemed to help the wound heal without scarring. Gattefossé's experience prompted explorations into the clinical use of essential oils and in 1937 led him to write a book in which he coined the term aromatherapy.

Today, some people use essential oils as part of a holistic approach to psychological and physical wellness. They believe that, since the olfactory neurons in our nose connect to the limbic system of our brain, natural aromas can evoke strong emotional reactions and produce positive psychological effects. People following this practice typically apply essential oils directly to the skin or carefully inhale them.

Essential oils are distilled from the leaves, stems, flowers, bark, and/or roots of a plant. Before being applied to the skin, the oils are diluted with a carrier oil such as sweet almond oil, apricot kernel oil, or grapeseed oil. Essential oils should never be taken orally since some are quite toxic. Additionally, they should be kept away from open flame due to their flammability. Essential oils are very expensive to produce due to the labor-intensive process and the amount of plant material required. For example, the peel from 1 ton of lemons yields only about 14 pounds of oil.

## Fragrance Industry

Electrical impulses travel from our nose to regions of our brain that control emotion, behavior, and memory, explaining why scents can cause specific emotional and behavioral responses. The fragrance industry tries to exploit this fact to full advantage. The American consumer is inundated with advertisements about perfumes and colognes. In fact, advertising and marketing campaigns account for most of the cost of perfumes. Television commercials often show handsome men and women wearing these fragrances in positive social situations. In addition to selling the fragrance itself, these commercials try to sell the positive image of the person wearing the fragrance. For example, a fragrance may be shown as sexy, mysterious, or alluring.

Manufacturers often add fragrance to personal care, personal hygiene, and household products to make the products more profitable and to cover up the smell of the product itself. Many soaps originally have a tallowy beef or coconut smell. To help boost the sales of their products, manufacturers spend a great deal of money on research to identify the fragrances that will please most of their target audience. In fact, marketing of these products sometimes focuses more on the fragrance of the product than on its performance.

Household products are marketed to make people think they have to scent the air in their homes to eliminate bad odors. Air fresheners sold as sprays, solids, and scented oils come in many floral, berry, citrus, and spicy scents. Although candles were once used primarily as a source of light, today most consumers buy candles to add fragrance to a room or to set a mood. Cleaners that were once available in only pine or lemon scent now come in a wide variety of scent options.

## Electronic Noses

Recent advances in technology have helped us begin to more fully capitalize on the importance of smell. In the mid-1980s, the Institute of Olfactory Research at the University of Warwick, Coventry, UK, developed a prototype electronic nose called the Warwick Nose. High-tech companies now make commercial versions of this electronic nose that essentially mimic functions of the human nose. These electronic noses have a sensor array or spectrometer that acts as the chemical-sensing system and an artificial neural network (ANN) that acts as the pattern-recognition system. The ANN works like a biological brain to identify chemicals and odors through a learning process. These instruments can be trained to detect and classify odors, vapors, and gases.

Currently, scientists are experimenting with the use of electronic noses to identify possible medical problems. By examining body odors from breath, wounds, and body fluids, electronic noses can help physicians diagnose diseases, monitor wound infections, track glucose levels in diabetics, and detect pathological conditions like tuberculosis. Electronic noses can evaluate a patient for infection within minutes, while a conventional bacterial culture takes 1–3 days.

Other applications for the electronic nose include environmental monitoring for air quality, natural gas or oil leaks, and toxic waste. The electronic nose can be used in the food industry for controlling fermentation, detecting rancidity, and inspecting and grading food by odor. Researchers are experimenting with a robotic lobster with an electronic nose for detecting TNT in hidden land and sea mines. Since we all have a unique body odor, perhaps some day our odor will be an alternative form of identification. Electronic noses could be used in security entry systems or to sniff out credit card fraud and fake IDs.

While electronic noses have limitations, they also have some significant advantages over the human nose. Electronic noses do not experience olfactory fatigue like human noses do, and their performance does not fluctuate based on health, mood, hormonal cycles, fatigue, working conditions, or other human factors.

## Development of the Stink Bomb

In 1998, as part of a nonlethal weapons program, the U.S. Department of Defense hired the Monell Chemical Senses Center in Philadelphia to research chemicals that could be used as malodorants in a stink bomb. The goal was to find a mixture of chemicals so universally repulsive that people of all cultures would smell the odor and have an overwhelming desire to flee. The Pentagon wanted to develop the foul odor to control riots, deter people from restricted areas, and drive away enemy troops. Use of the odor would be far less dangerous than rubber bullets, batons, tear gas, or pepper spray.

Researchers used test subjects from five ethnic groups and focused on odors with biological origins such as body odors, vomit, human waste, rotting flesh, burnt hair, and decaying garbage. Surprisingly, some test subjects showed particular tolerance to vomit and burnt hair. Two universally repulsive odors were those from human feces and from a mixture of decaying flesh and garbage. To chemically synthesize the odor of feces, researchers mixed a chemical called skatole with organic sulfur compounds and fatty acids like those typically found in vomit, foot sweat, and strong cheese. (See Figure 1-5.) Synthesis of the rotting organic material odor was achieved by combining a variety of organosulfur compounds like mercaptoacetic acid. Since people can become acclimated to individual odors, mixing these two odors together creates an odor that lasts longer and is more powerful than either one by itself. The result is a stink bomb causing subjects to flee due to shortness of breath, rapid heart rate, and nausea.

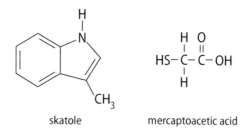

Figure 1-5: Some stink bomb ingredients

## ● Body Scents

Similar to a fingerprint, each person has a unique body scent. To some extent, humans can distinguish each other by scent. For example, a breast-fed newborn baby is attracted to a piece of cotton rubbed against its mother's neck but will turn away from cotton swabbed against a stranger's neck. Mothers also show an amazing ability to distinguish their newborn babies by smell for a short time after birth. Blindfolded mothers can even identify clothes their babies have worn recently by smelling the clothing items.

Some body odors tend to be characteristic of a person's genotype, including gender and race, or of cultural habits, such as diet and hygiene practices. A study of college students indicated that odor was a good detector of gender: musk was generally associated with males and "sweet" with females. Often the food we eat provides body odors that last an extended time, and studies show that persons eating large quantities of some foods (such as cabbage, garlic, curry and other spices, butter and other fats, or fish) often exhibit characteristic odors of these foods.

Sweat glands are often blamed for generating body odors. In reality, sweat itself does not have an unpleasant odor. However, when sweat collects in warm, moist places where it does not quickly evaporate (such as under the arms or inside shoes), bacteria native to these regions break the sweat down into compounds with unpleasant odors we typically associate with sweat.

Sweat glands are one of the major body cooling systems—as sweat evaporates from the surface of the skin, heat is removed from the skin and the body is cooled. There are two types of sweat glands, apocrine and eccrine. (See Figure 1-6.) These glands differ in body location, sweat composition, age at which they become active, and conditions that trigger their activity.

The genetic makeup of a person determines the amount and distribution of apocrine sweat glands, although the glands are generally concentrated under the arms and in the genital region. Apocrine glands are modified sebaceous glands and, just as sebaceous glands secrete sebum at the pit of the hair follicles to keep hair and skin lubricated, apocrine glands secrete a viscous, sometimes milky or yellowish, oily "scent substance" consisting of proteins, lipids, and steroids. Apocrine glands are small until puberty, when their growth is stimulated by hormonal changes in the body. These glands are emotionally triggered and do not play a role in temperature regulation.

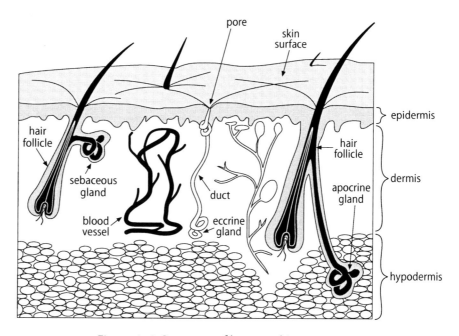

Figure 1-6: Structure of human skin

Eccrine sweat glands are located all over the body, although they are more abundant in the palms of hands, soles of feet, and forehead. These glands are long, coiled tubes of cells that connect to pores on the skin's outer surface by long ducts. Eccrine glands secrete mostly water and salt. They are active from birth and are triggered by exercise, hot temperatures, and emotional states such as fright, nervousness, embarrassment, anger, or anxiety.

Apocrine, eccrine, and sebaceous glands can contribute to body odor. Apocrine sweat contains organic compounds that are decomposed or modified by bacteria living on the skin, resulting in the production of odorous molecules. Sweat from eccrine glands provides the moisture necessary for the bacteria to grow. Sebaceous glands are not sweat glands, but they secrete an oily sebum that some scientists believe contains components that may be odorant precursors.

# Chapter 2
# The Nose Knows—Exploring the Sense of Smell

Chapter 2 Overview .................................................................................................. 18
National Science Education Standards ................................................................ 18
Cross-Curricular Integration.................................................................................. 19

Section 1 Background: How Your Sense of Smell Works.................................... 21
Section 1 Experiment: Identifying Odors ............................................................. 23
Instructor Notes for Section 1............................................................................... 25

Section 2 Background: Knowing Your Sense of Smell ....................................... 30
Section 2 Experiment: Get a Whiff of This ........................................................... 32
Instructor Notes for Section 2............................................................................... 34

Section 3 Background: Molecules with a Message ............................................ 36
Section 3 Experiment: Molecules You Can Smell ............................................... 38
Instructor Notes for Section 3............................................................................... 40

# Chapter 2 Overview

**Key Science Topics**
- physiology and chemistry of odor perception
- chemical nature of odor-producing molecules

Smell and our other senses work together to enable us to observe and interpret the world around us. Odors can alert us to dangers like fires and natural gas leaks. Our noses also help us to taste and enjoy our favorite foods. Our sense of smell can even lead us to recognize places and recall special memories like swimming in the ocean or hiking in the woods.

How does our brain recognize an odor? Section 1 of this chapter introduces the physiology and chemistry of smelling. The Section 1 experiment asks students to smell unknown substances and use their previous knowledge about odors to identify the unknown odors. Students learn that, without using other senses like seeing and tasting, this task can be challenging. Section 2 addresses differences in individuals' abilities to detect odors. In the Section 2 experiment, students take a standard clinical test to assess their sense of smell. They learn that the ability to smell differs greatly among individuals. Section 3 of this chapter explores the common characteristics of odor-causing molecules. Students identify unknown substances by odor and guess the familiar scent produced when the unknown odors are blended together.

## National Science Education Standards
This chapter addresses the National Science Education Content Standards for grades 5–8 and 9–12 as described in the following lists.

### Section 1
*Science as Inquiry:*
Abilities Necessary to Do Scientific Inquiry
- Students predict the identities of household chemicals based on odor alone.

*Life Science:*
Structure and Function in Living Systems
- Specialized cells perform specialized functions in multicellular organisms. Each type of cell, tissue, and organ has a distinct structure and set of functions that serve the organism as a whole. Students learn how their sense of smell works and about the specialized cells involved in the process.

### Section 2
*Science as Inquiry:*
Abilities Necessary to Do Scientific Inquiry
- Students conduct an investigation to assess their sense of smell.
- Students develop explanations for differences in their results.

*Life Science:*
Structure and Function in Living Systems
- Disease is a breakdown in structures or functions of an organism. Some diseases are the result of intrinsic failure of the system and others are the result of damage by infection. Through the background reading, students learn about anosmia, the condition of being unable to smell.

*Science in Personal and Social Perspectives:*
Personal Health
- The potential for accidents and the existence of hazards imposes the need for injury prevention. Through the background reading, students learn of different hazards that people who suffer from anosmia may face and the need for clinical tests to diagnose the condition.

## Section 3
*Science as Inquiry:*
Abilities Necessary to Do Scientific Inquiry
- Students predict the identities of unknown substances based on odor and guess the familiar scent produced when the unknown odors are blended together.

*Physical Science:*
Properties and Changes of Properties in Matter
- A substance has characteristic properties. Students learn the general properties of odor-causing molecules.

## Cross-Curricular Integration
The experiments in this chapter can be integrated with other areas of the curriculum to emphasize the relationships between subjects. Some ideas for cross-curricular integration are listed below.

### Language Arts
- Have students select a smell associated with certain memories and write an essay about those memories.
- Have students write descriptive paragraphs that paint pictures of places by describing smells. Topics could include the county fair, a summer garden, or a grandparent's house at a favorite holiday time.

### Life Science
- Study how different animals use their abilities to detect odors and the roles that odors play in the daily lives of animals (such as locating food, defense, and mating).

## Social Studies

- The odors of foods from other cultures may be difficult for students to identify. Have students cook and smell unfamiliar foods from different countries—for example, curry (India), chili peppers (Mexico), or fish sauce (Vietnam).
- Have students research careers that require using the sense of smell.

# ● Section 1 Background: How Your Sense of Smell Works

The inside of your nose is called the nasal cavity. (See Figure 2-1.) This cavity contains long nerve cells called olfactory neurons, which are specially designed for smelling. Sources vary widely on the number of olfactory neurons that humans have; some indicate that humans have 5–50 million olfactory neurons. The upper part of each olfactory neuron connects to an olfactory bulb at the base of the brain. The other end of each neuron branches out into about eight cilia (hairlike fingers) that reach into the nasal cavity. Mucus from olfactory (Bowman's) glands forms a moist layer around the cilia.

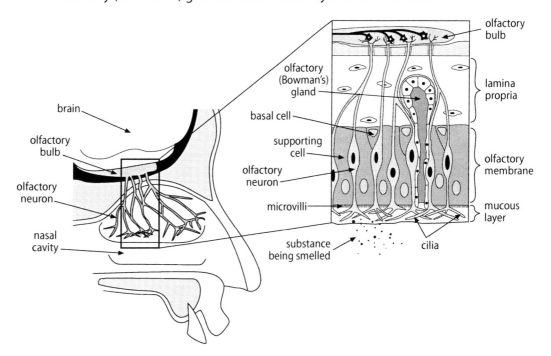

Figure 2-1: Cross-section of the face

When you breathe through your nose, air carrying molecules of odorous substances swirls around in your nasal cavity. Smelling begins when odorous molecules dissolve in the mucous layer around the cilia. These odorous molecules have specific sizes and shapes that fit into and interact with odor receptors on the cilia. In the process, the olfactory neuron produces an electrical signal. This signal travels through the olfactory bulb at the base of the brain and into several areas of the brain, where it is interpreted as odor.

The nose contains 400 or more different types of olfactory receptors, which together can recognize about 10,000 distinct odors. Each receptor may be able to recognize more than one type of odorous molecule, and the brain may use a complex code of signals from several neurons to identify an odor.

# Section 1 Background: How Your Sense of Smell Works

Our sense of smell also helps in tasting food. Since the nasal cavity and the oral cavity (mouth) are connected in the back of the throat, odor information combines with taste information as we eat to create the sensation of flavor. That is why a person with a stuffy nose sometimes has trouble tasting food.

*Oklahoma, May 4, 1999—Urban Search and Rescue teams worked to recover missing persons after a devastating tornado.*

**Without a Trace?** ▶ Dogs are often used in police work to search for criminal suspects and missing persons. Dogs can track human scents far better than humans or existing technology. Dogs have an acute sense of smell due to their extraordinary number of olfactory neurons—a number far exceeding that of humans.

So how do dogs pick out one human scent from another? Human scent is composed of a mixture of skin cells, perspiration, skin oils, and gases. Each person has a nearly unique mixture of these components that helps to create a unique scent signature, like a fingerprint.

Scent trails are formed by the skin cells, called skin rafts, that fall from our bodies at a rate of about 40,000 cells per minute. These rafts can be carried upward and away from the body by air currents. Temperature, humidity, sun exposure, and wind determine how long the scent can be detected. Dogs must be called in as quickly as possible to search a scene.

# ● Section 1 Experiment: Identifying Odors

How many common substances can you identify by odor alone?

## Materials

Per class
- containers holding different substances

## Procedure

**Caution:** *Whenever you smell an unknown substance, use the wafting technique to protect yourself. If the odor is unpleasant, do not continue to smell it. Typically, unpleasant odors are not good for you.*

❶ Containers holding different substances will be passed around the classroom. Do not open or tip over these containers. As each container reaches you, gently waft the vapors toward you to smell the odor coming from the holes in the top of the lid. (See Figure 2-2.)

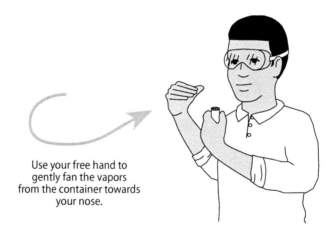

*Figure 2-2: Wafting technique*

❷ After you waft each unknown substance, record a brief description of its odor on the Data Table. Use as many descriptive words as possible, such as fruity, citrus, overpowering, cleaning agent, antiseptic, and deodorizer. You can use words that may only be familiar to you, such as Grandpa's house or similar memory-based descriptors.

❸ After completing step 2 for each unknown substance, work in small groups as assigned by your teacher. Attempt to identify each substance based on your written description and record your predictions in the Data Table.

Section 1 Experiment: Identifying Odors

④ Your teacher will give you a list (in random order) of the actual substances used in the experiment. Use this list to refine your predictions. Record your new predictions in the Data Table.

⑤ In the last column of the Data Table, record the actual identity of the substance in each container based on information provided by your teacher.

| Data Table for Identifying Odors | | | | |
|---|---|---|---|---|
| Number of Unknown | Step 2: Description of Odor | Step 3: Predicted Identity Based on Observations Only | Step 4: Predicted Identity Using List of Possibilities | Step 5: Actual Identity Provided by Teacher |
| 1 | | | | |
| 2 | | | | |
| 3 | | | | |
| 4 | | | | |
| 5 | | | | |
| 6 | | | | |
| 7 | | | | |
| 8 | | | | |
| 9 | | | | |

**Question 1:** *Which prediction changed most dramatically between steps 3 and 4? Give a reason why this substance was particularly difficult to identify initially.*

**Question 2:** *Which substance was the easiest to identify? Why?*

**Question 3:** *If you misidentified some odors, why do you think that may have happened? What other clues do we use to identify a particular odor?*

**Question 4:** *Explain how the process of identifying the unknown substances would have been different if you'd been given the list of actual substances used at the beginning of the experiment rather than near the end.*

# Instructor Notes for Section 1

In this experiment, students try to identify common household chemicals by odor alone. After describing the odors in the containers, students make predictions about the identities of the substances; first, without knowing the possible substances, and second, after receiving a scrambled list of the actual substances.

## Time Required

Setup: 10 minutes (plus 20 minutes the first time the experiment is done)
Procedure: 20–25 minutes
Cleanup: 5 minutes

## Materials

For Getting Ready
Per class

- film canisters or other small opaque containers with plastic lids (1 container for each household chemical used)
- extra plastic lids (1 for each container used)
- cotton balls
- pushpin
- (optional) zipper-type freezer bags
- water

 *Use distilled water if tap water has a high sulfur content or is highly chlorinated.*

- some or all of the following common household chemicals:
    - vinegar
    - musk-scented aftershave or cologne
    - peppermint oil
    - orange extract
    - banana extract
    - floral air freshener
    - vanilla
    - Noxema® skin cream

For the Procedure
Per class

- containers holding different substances (prepared in Getting Ready)

Chapter 2: The Nose Knows—Exploring the Sense of Smell

**Instructor Notes for Section 1**

## Safety and Disposal

Warn students to use the wafting technique when smelling unknown substances. Avoid dangerous household substances such as bleach, ammonia, or battery acid. Have students wear goggles to reinforce proper laboratory safety techniques.

Dump cotton balls and solid waste into plastic bags, tie off, and discard.

## Getting Ready

❶ Number each container (not the lid).

❷ Use water as one of the test substances and select some or all of the common household chemicals in the Materials list for the remaining test substances. For each liquid sample, place a cotton ball in the container and pour a small amount of liquid onto the cotton ball. Do not saturate the cotton ball because the liquid might spill out of the container. For each solid sample, put a small amount into the container. Keep a key of the substance used in each numbered container. This key will be used for step 5 of the experiment. Also prepare an unnumbered, scrambled list of the test substances to be used for step 4 of the experiment.

❸ Close the prepared containers with the solid plastic lids. To avoid cross-contaminating odors, store the containers separately (preferably in zipper-type freezer bags).

❹ Poke 10–20 holes in the extra lids with a pushpin to prepare porous lids that will replace the solid lids just before the experiment.

## Procedure Notes

When you are ready to begin the experiment, replace the solid lids on the containers with the porous lids. Do not allow students to open the containers. Instruct students to identify the odor by first wafting the odor from the holes in the top of each container. If an odor is not detected, the container can be waved closer to the nose.

Extend the learning in the experiment by discussing wafting and other techniques for smelling known and unknown substances. Wafting is the correct technique for smelling substances in the lab. However, because this technique fills a room with odors rather quickly, scent panels in industry do not use the wafting technique. Rather, these panels wave open containers under their noses because this method causes less contamination of the room air. Note that before a scent panel receives test substances, the concentration of odor chemicals are adjusted to a comfortable, detectable level for the scent panel, so the possibility of harmful overexposure is eliminated.

If using the containers with more than one class, replace the porous lids with the solid lids between classes. The containers can be reused in subsequent years.

## Answers and Observations

**Question 1:** *Which prediction changed most dramatically between steps 3 and 4? Give a reason why this substance was particularly difficult to identify initially.*

Answers will depend on the students' sense of smell and prior experience with the unknown test substances. Students may indicate that water was the most difficult to identify initially, since it has little or no odor. However, once students are given the list of possible substances, they can identify water as the substance with little or no odor.

**Question 2:** *Which substance was the easiest to identify? Why?*

Answers will depend on the students' sense of smell and prior experience with the unknown test substances.

**Question 3:** *If you misidentified some odors, why do you think that may have happened? What other clues do we use to identify a particular odor?*

Students may misidentify odors for many reasons, including a student's unfamiliarity with the odor, the odor's close relationship to another odor, or a student's inability to smell clearly due to a cold, allergies, or other condition. Additionally, people usually use visual and/or taste clues to help identify odors.

**Question 4:** *Explain how the process of identifying the unknown substances would have been different if you'd been given the list of actual substances used at the beginning of the experiment rather than near the end.*

Most students will say that having the list of actual substances used in the experiment would have made the identification process easier. They would have been able to use the process of elimination.

## Explanation

This experiment challenges students to identity substances that have characteristic odors without any clues other than smell. Without being able to see or taste the substances, this task can be difficult. The label we apply to a given odor is a learned response. For example, we are taught that the citrus-like fruity smell of oranges, orange juice, and orange blossoms is an orange smell. But without the visual or taste clues, students usually can only describe the odor as smelling like citrus.

**Instructor Notes for Section 1**

*Overhead Master: Cross-Section of the Face*

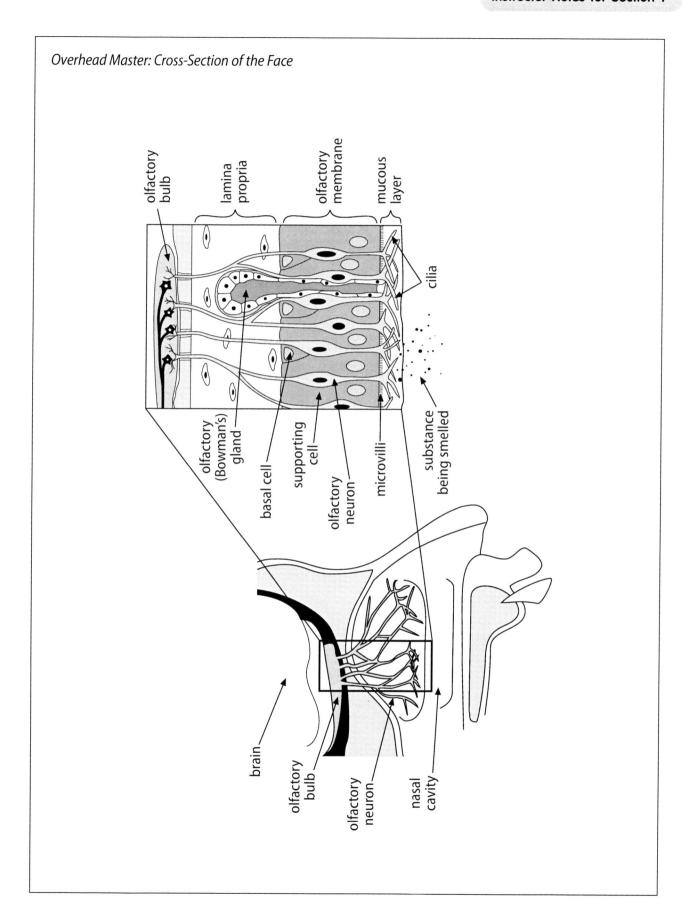

Chapter 2: The Nose Knows—Exploring the Sense of Smell

# Section 2 Background: Knowing Your Sense of Smell

Not everyone detects a scent to the same degree. Some people can smell more kinds of odors or can detect fainter odors than other people. Generally, women can detect a fainter scent more easily than men and nonsmokers are better at detecting scents than smokers. One survey found that over 2 million Americans said they can't smell anything at all! Scientists have found that the sense of smell declines after age 60.

Recent smell research has resulted in several standard tests to measure a person's sense of smell and to help doctors diagnose smelling disorders. The Nasal Dysfunction Clinic at the University of California at San Diego developed a simple smell test called the alcohol sniff test (AST). This test measures how close an alcohol pad must be to a person's nose before the alcohol is smelled. The University of Pennsylvania Medical Center developed a smell test in which patients smell and try to identify 40 different test odors.

People with a loss of smell have a condition called anosmia. A common cause of anosmia is rhinitis, which is inflammation of the nasal membranes. If the

*George Aldrich at work sniffing for NASA.*

**Getting Paid to Smell** ▶ George Aldrich has a job that really stinks sometimes. He works as a professional sniffer at NASA's Johnson Space Center, White Sands Test Facility, Las Cruces, NM. Researchers have learned that odors can be a problem inside space capsules because of the confined area and high temperatures, especially during re-entry. Aldrich is in charge of 20–25 people who smell everything that goes into the living quarters of space shuttles and the International Space Station. Items such as pens, shaving cream, toothpaste, socks, and shoes are placed in separate sealed containers and heated to 120°F (about 49°C) for three days. Samples of air from each container are tested for toxicity. Nontoxic samples are tested further by a panel of five people who smell the samples to determine if the odors will be offensive to astronauts. Each sniffer rates the odors on a scale of 0 (undetectable) to 4 (revolting). Samples with an average score above 2.4 fail the sniff test.

Aldrich has participated in over 750 smell panels, which he refers to as "smell missions." How do you become a professional sniffer for NASA? All candidates must take a special physical because sniffers can't suffer from allergies or respiratory problems that irritate the nasal passages. Once a sniffer is on the job, his or her nose has to be certified every four months by passing a special test.

Being a professional sniffer is pretty serious business. Since a lot of money goes into each space mission, NASA doesn't want to jeopardize a mission because of a foul smell.

nasal membranes are swollen and inflamed, air can't reach the olfactory neurons and a person can't smell. This inflammation can be caused by many things, including head colds, infections, allergens, cigarette smoke, and air pollutants. A deviated nasal septum, crooked nose, or nasal tumor can also block air from reaching the olfactory neurons and cause anosmia. Anosmia can also be caused by damaged olfactory neurons due to viral infections, head injury, nutritional deficiencies, old age, and mental disorders. Depending on the cause, anosmia may be treated with antibiotics, decongestants, antihistamines, steroids, trace metals, or surgery.

Anosmics should be aware of their condition so that they can take certain safety precautions. For example, since they can't smell smoke, smoke detectors should be installed in the kitchen, in every room with a fireplace, and in every room where the person might sleep. If the house is heated with propane or natural gas, special gas detectors should be added to the house since the person wouldn't be able to smell a gas leak. People with anosmia should also take special care when selecting foods to eat, especially leftovers, since they are unable to smell spoiled food and are thus more at risk of food poisoning.

# Section 2 Experiment: Get a Whiff of This

How keen is your sense of smell? Take the alcohol sniff test (AST) and find out.

## Materials
Per pair of students
- 2 sheets of paper
- 2 individually packaged, disposable 70% isopropyl alcohol pads
- metric ruler

## Safety and Disposal
Alcohols are flammable and should be kept away from open flames.

## Procedure

1. Tear the top of the wrapper away from one of the alcohol pads so that one-half to one-third of the alcohol pad is exposed. Sniff the pad to learn what the alcohol smells like.

2. Fold one sheet of paper in fourths lengthwise. Label one end "top."

3. Test your sense of smell:

   a. Hold the top of the paper against the upper lip so that the paper touches the bottom of the nose and hangs down and out at about a 15° angle from the body. (See Figure 2-3.)

   b. Close your eyes and breathe normally. Your partner should hold the alcohol pad at the bottom edge of the paper. Each time you exhale, the partner should move the alcohol pad up the paper about 1 cm closer to your nose. Tell your partner when you first detect the alcohol odor. The partner should draw a line on the paper along the top of the alcohol pad.

   c. Use a ruler to measure how far the top of the alcohol pad was from your nose when the alcohol odor was first detected. Record the result in the Data Table.

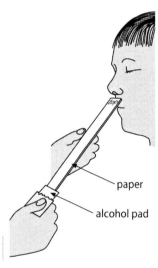

*Figure 2-3: AST setup*

4. Repeat step 3 with the same student and the same alcohol pad. Turn the paper over for this second test.

**Question 1:** *Was there a difference between your first and second measurements? Why or why not?*

# Section 2 Experiment: Get a Whiff of This

**❺** Now switch roles to test the smelling ability of the second student. Using a new alcohol pad and a new piece of paper, repeat steps 1–4 and record the results on the Data Table.

**Question 2:** *Was there a difference between your measurements and the measurements of your partner? Why or why not?*

**❻** Collect class data and compare the results of males versus females.

**Question 3:** *How does odor detection sensitivity compare between males and females in your class?*

| Data Table for Smell Test | | | |
|---|---|---|---|
| Student | Male or Female | Measurement for Trial 1 in cm | Measurement for Trial 2 in cm |
| 1 | | | |
| 2 | | | |

# Instructor Notes for Section 2

Students perform a threshold clinical test called the alcohol sniff test (AST) to assess their sense of smell. Learning about nasal dysfunction helps students realize that not all people have the same sense of smell.

## Time Required
Setup: 5 minutes
Procedure: 15 minutes
Cleanup: 5 minutes

## Materials
Per pair of students
- 2 sheets of paper
- 2 individually packaged, disposable 70% isopropyl alcohol pads

 *Alcohol pads are typically used in hospitals and doctors' offices and can be purchased in many pharmacies.*

- metric ruler

## Safety and Disposal
Alcohols are flammable and should be kept away from open flames.

## Answers and Observations

**Question 1:** *Was there a difference between your first and second measurements? Why or why not?*

Answers will vary depending on the data. Some variation between trials may occur due to olfactory fatigue or inaccurate measurements.

**Question 2:** *Was there a difference between your measurements and the measurements of your partner? Why or why not?*

Answers will vary depending on the data. Some variation is to be expected because individuals' senses of smell vary widely.

**Question 3:** *How does odor detection sensitivity compare between males and females in your class?*

Answers will vary depending on the data. Research suggests that females are more sensitive to odors than males.

**Instructor Notes for Section 2**

## Explanation

Remind students that this experiment is only a simulation of a clinical procedure, so the results should not be considered as a medical diagnosis. People with a normal sense of smell can usually detect the alcohol odor when the alcohol pad is held at a distance of 20 cm or greater from the nose. People having hyposmia (a decreased ability to smell) detect the alcohol odor at 2–20 cm. Anosmia patients can't smell the alcohol odor at all, although they may be able to feel coolness on their nose due to evaporation of the alcohol.

Remind the students that hyposmia and anosmia are sometimes temporary conditions. For example, a student may have nasal congestion on the day of the experiment. Discuss possible causes and available treatments for hyposmia and anosmia, as outlined in the Background portion of this section.

# Section 3 Background: Molecules with a Message

Some molecules produce characteristic odors and others do not. Although scientists do not agree on exactly what properties cause this phenomenon, studies of odor-causing molecules have lead to several general conclusions.

- Most odor-causing molecules belong to the class of chemicals known as organic chemicals. Organic chemicals are molecules that have carbon atoms bonded together in straight or branching chains or rings. (See Figure 2-4.)

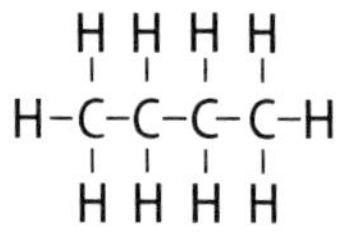

shows carbon atoms in a chain

shorthand notation

n-butane

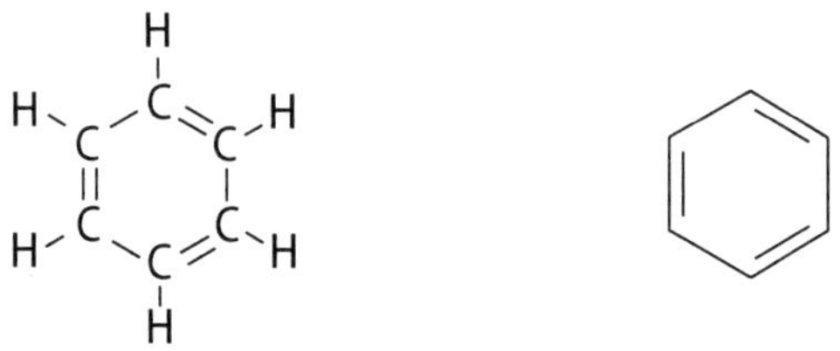

shows carbon atoms in a ring

shorthand notation

benzene

*Figure 2-4: Two organic molecules*

- Molecules that produce distinct odors are often small to intermediate in size.

- Molecules that produce the strongest odors are often soluble in both fat and water.

- Humans can smell even very small amounts of some odors. For example, mercaptans produce a skunk odor that can be smelled at a concentration of 1 part per 50 billion. (See Figure 2-5.)

$$CH_3-CH_2-S-H$$

*Figure 2-5: Ethyl mercaptan*

- Even relatively small changes in the structure of a molecule can drastically change its odor. Can you find the differences in the molecules in Figure 2-6?

## Section 3 Background: Molecules with a Message

**benzaldehyde**
(strong odor of bitter almonds)

**benzoic acid**
(nearly odorless)

**vanillin**
(odor of vanilla)

**ethyl vanillin**
(3 times stronger vanilla odor than vanillin)

*Figure 2-6: Minor molecular changes resulting in major odor differences*

*Substances added to natural gas give it odor.*

**Adding Odor to Natural Gas ▶** If you have a gas stove in your home, you may have noticed that the unlit gas has an unpleasant odor similar to rotten eggs. But the natural gas used for cooking and for heating homes does not start out with a bad smell. Processed natural gas is actually odorless; the odor is added as a safety precaution so you will be alerted to a gas leak.

Raw natural gas comes from porous rocks found deep underground and consists of about 75% methane ($CH_4$), 15% ethane ($C_2H_6$), and 5% other hydrocarbons such as propane ($C_3H_8$) and butane ($C_4H_{10}$). Before being used, most raw natural gas is processed to separate out the heavier hydrocarbon liquids and to remove contaminants. The remaining processed natural gas is over 90% pure methane, which is odorless, colorless, tasteless, and very flammable. Methane can cause suffocation by reducing the amount of oxygen available for a person to inhale.

Manufacturers add odor to processed natural gas using substances such as mercaptans, which smell like rotten eggs. Very little mercaptan, only a few parts per million, is required to give natural gas an odor. This amount is enough to be detected by a person with a normal sense of smell when the natural gas is at 0.5% concentration in air.

**Chapter 2: The Nose Knows—Exploring the Sense of Smell**

# Section 3 Experiment: Molecules You Can Smell

Some molecules that produce odors and flavors can be found in the foods and beverages that we consume. Can you identify the odors of some unknown substances and the familiar scent that they produce when their odors are blended together?

## Materials

Per group
- 2 sets of test tubes
- piece of waxed paper

## Safety and Disposal

Never taste these or any substances used in the laboratory.

## Procedure

1. Each group will be given a set of four unknown substances labeled A–D. Each test tube contains a different unknown substance. Remove the paper strip from test tube A. Each group member should wave strip A under his or her nose. Record the description of the odor in the Data Table. After each member has smelled strip A, place it on a piece of waxed paper for later use. Repeat this procedure with substances B–D, one at a time.

2. Each group will be given a second set of identical substances. This time the names of the substances are provided. Smell the substances as in step 1 and place the strips on the waxed paper. Try to match the unknown substances with the known substances. Record your results in the Data Table.

3. Your teacher will tell you the actual identities of the unknown substances from step 1. Record the identities in your Data Table and check whether you were able to successfully identify the unknowns in either step 1 or step 2.

4. Hold the set of unknown strips or the known strips in a fan arrangement near your nose. Gently wave the fan to combine odors and record your description of the odor. Don't reveal the secret until all students have completed this step.

   **Question 1:** *What familiar scent do you think the odor mixture smells like?*

5. In groups, discuss and compare the guesses of each person.

   **Question 2:** *How many people had the same guess as you? What was the most common guess for the scent?*

## Section 3 Experiment: Molecules You Can Smell

| | Data Table for Odor Identification | | |
|---|---|---|---|
| Substance | Step 1: Description of Odor | Step 2: Predicted Identity Based on Known Substances | Step 3: Identity Provided by Teacher |
| A | | | |
| B | | | |
| C | | | |
| D | | | |
| A–D | Description of combined scent: | | |

# Instructor Notes for Section 3

Students learn that some ingredients found in food and drink contain molecules that produce odor and flavor. After identifying unknown substances by odor, students guess the familiar scent produced when the unknown odors are blended together.

## Time Required
Setup: 10 minutes
Procedure: 20–25 minutes
Cleanup: 5 minutes

## Materials
For Getting Ready
Per class
- marker
- filter paper or chromatography paper
- scissors or paper cutter
- lime oil
- orange oil
- lemon oil
- cinnamon oil

 *These oils are available from a grocery store or pharmacy. They are typically used in candy making.*

Per group
- 8 test tubes with stoppers
- 2 beakers or test tube racks to hold the test tubes

For the Procedure
Per group
- 2 sets of test tubes
- piece of waxed paper

## Safety and Disposal
Warn students never to taste substances used in the laboratory.

## Getting Ready
For each group of students, place four test tubes in a beaker or test tube rack. Use a marker to label the test tubes A, B, C, and D. Cut filter paper or

chromatography paper into thin strips that fit inside the test tubes, making enough strips so that each group gets one strip from each test tube. Label the top of each strip of paper A, B, C, or D. Place one or two drops of lime oil in test tube A, one or two drops of orange oil in B, one or two drops of lemon oil in C, and one or two drops of cinnamon oil in D. Place a paper strip in each test tube. Stopper the test tubes.

Make a second set of test tubes for each group of students. Repeat the above procedure, but label the test tubes and strips of paper with the substance names.

Since the oils are concentrated, we suggest you use the smallest amounts of oil to just moisten the edge of the paper strips. If one component is more overpowering than the other three, the combined scent may be difficult to identify.

## Procedure Notes

Encourage the students to be as specific as possible when describing the odors. When an odor is fruity or citrus-like, students should try to name the specific fruit.

Do not distribute the second set of test tubes until after step 1 to prevent students from prematurely learning the identity of the known substances.

## Answers and Observations

**Question 1:** *What familiar scent do you think the odor mixture smells like?*

The odor mixture smells like the cola flavor used in soft drinks.

**Question 2:** *How many people had the same guess as you? What was the most common guess for the scent?*

Answers will vary depending on the data.

## Explanation

A combination of several aroma compounds may produce a recognizable odor or flavor. This is often what makes one brand of consumer product preferred to another product. Cola is an example of such a combination and is also known as a fantasy flavor since it does not mimic a natural scent or flavor. Citrus fruit flavors and spice flavors are combined to produce the unique cola flavor. Although manufacturers of cola-flavored soft drinks want to keep their exact

### Instructor Notes for Section 3

cola formulas a secret, we know that cola usually contains cola nut extract; caramel color; one or more of grapefruit oil, lemon oil, lime oil, and orange oil; and one or more of cinnamon oil, clove oil, coriander oil, neroli oil, nutmeg oil, and ginger oil.

In the experiment, four common aroma compounds are combined to produce the fantasy scent of cola.

*Overhead Master: Two Organic Molecules*

$$\text{H-C-C-C-C-H structure with H's} \quad CH_3-CH_2-CH_2-CH_3$$

shows carbon atoms in a chain ——— shorthand notation

n-butane

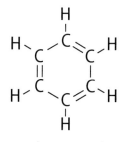

shows carbon atoms in a ring ——— shorthand notation

benzene

*Overhead Master: Minor Molecular Changes Resulting in Major Odor Differences*

**benzaldehyde**
(strong odor of bitter almonds)

**benzoic acid**
(nearly odorless)

**vanillin**
(odor of vanilla)

**ethyl vanillin**
(3 times stronger vanilla odor than vanillin)

# Chapter 3
## The Origins of Body Odor

Chapter 3 Overview ................................................................................................ 46

National Science Education Standards ................................................................ 46

Cross-Curricular Integration ................................................................................. 48

Section 1 Background: Your Unique Scent ........................................................... 49

Section 1 Experiment: Scent Detectors ................................................................ 51

Instructor Notes for Section 1 .............................................................................. 53

Section 2 Background: Everything You Always Wanted to Know About Sweat ... 56

Section 2 Experiment: Using a Sweat Meter ........................................................ 58

Instructor Notes for Section 2 .............................................................................. 59

Section 3 Background: No Sweat .......................................................................... 62

Section 3 Experiment: Sweating Is Cool .............................................................. 63

Instructor Notes for Section 3 .............................................................................. 65

# Chapter 3 Overview

**Key Science Topics**
- factors influencing an individual's unique scent
- eccrine sweat
- apocrine sweat
- bacterial decomposition of sweat
- physiological purpose of sweat

Teenagers go through many emotional and physical changes during their transition from childhood to adulthood. Hormones typically fluctuate during this time, often taking teens on an emotional roller coaster ride. Through it all, teen bodies undergo many physical changes that require getting used to. Understanding these physical changes is the first step in accepting them.

This chapter deals primarily with the science of sweat. Section 1 of this chapter helps students discover that we all have a unique scent based on our genetics, gender, and diet. In the Section 1 experiment, students become aware of the scents of their own hands and guess the gender and identity of volunteers based on hand scent. Section 2 deals with the science behind sweating. The Section 2 experiment explores the effects of sweaty hands on a toy cellophane fish. Section 3 covers the cooling effects of sweat. In the Section 3 experiment, students simulate this cooling effect.

## National Science Education Standards
This chapter addresses the National Science Education Content Standards for grades 5–8 and 9–12 as described in the following lists.

### Section 1
*Science as Inquiry:*
Abilities Necessary to Do Scientific Inquiry
- Students conduct an investigation to see if they can detect the gender and identity of a person based on hand scent.
- Students use appropriate tools and techniques to gather, analyze, and interpret data. Students collect the data in a table and use this table to analyze the results.
- Students use what they have learned in this experiment to explain the research procedures used in a University of California body odor study.

### Section 2
*Science as Inquiry:*
Abilities Necessary to Do Scientific Inquiry
- Students design and conduct an investigation of the factors that cause a cellophane fish to curl when placed in the palm of the hand.
- Students develop predictions and explanations for the factors that cause the cellophane fish to curl.

Chapter 3 Overview

Understandings about Scientific Inquiry
- Different kinds of questions suggest different kinds of scientific investigations. Some investigations involve observing and describing objects. Students design their own experiment to determine whether physical activity affects the behavior of the fish.

*Life Science:*
Structure and Function in Living Systems
- The human organism has systems for digestion, respiration, reproduction, circulation, excretion, movement, control and coordination, and protection from disease. Students examine the production of sweat, which helps the body regulate temperature.

## Section 3

*Science as Inquiry:*
Abilities Necessary to Do Scientific Inquiry
- Students conduct an investigation into the temperature changes that occur as various liquids evaporate.
- Based on evidence gained through this experiment, students develop explanations for how sweating cools the body.
- Students use mathematics to calculate the changes in temperature as various liquids evaporate.

*Physical Science:*
Properties and Changes of Properties in Matter
- A substance has characteristic properties, such as density, boiling point, and solubility—all of which are independent of the amount of the sample. Students examine evaporation time and resulting temperature change as various liquids evaporate.

Transfer of Energy
- Heat moves in predictable ways, flowing from warmer objects to cooler ones, until both reach the same temperature. Students learn that sweat evaporation carries heat energy away from the skin.

*Life Science:*
Structure and Function in Living Systems
- The human organism has systems for digestion, respiration, reproduction, circulation, excretion, movement, control and coordination, and protection from disease. Students investigate temperature change caused by the

Chapter 3 Overview

evaporation of various liquids and relate these findings to how sweat cools the body and helps to regulate body temperature.

## Cross-Curricular Integration

The experiments in this chapter can be integrated with other areas of the curriculum to emphasize the relationships between subjects. Some ideas for cross-curricular integration are listed below.

### Life Science

- Research the science behind the term "cold sweat."
- Research to find out which animals sweat and which ones don't and how this may affect them.
- Research ways that animals keep themselves cool if their sweat glands are not highly active. For example, cats have sweat glands in the pads of their paws, but they also pant like dogs to regulate body temperature. Elephants, who only have sweat glands near their toenails, stay cool using methods such as flapping their ears.

# ● Section 1 Background: Your Unique Scent

We scrub our skin, we shampoo our hair, we use deodorants and perfumes; yet we all have our own unique personal odor. Even babies are aware of their mothers' special smell. In one study, babies turned toward cotton pieces that were rubbed on their mothers' necks, but they turned away from cotton rubbed on strangers' necks.

Specially trained dogs can track a missing person by matching the person's scent on clothing or other belongings to the scent left by the person. These dogs sniff the missing person's belongings to become trained on the scent and then they follow the scent trail. Humans are not nearly as good at tracking by scent, or we would probably use humans for this job instead of dogs!

Body odor is related to gender, genetics, race, cultural habits such as diet and hygiene practices, and age-related life cycle phases such as puberty. In several studies, participants ranging from nine years old to college age could generally determine by smell alone whether a T-shirt had been worn by a male or female (when the wearer had not used soaps, deodorants, or perfumes for 24 hours before wearing the shirts).

How do our bodies create scent? We have two types of sweat glands—apocrine and eccrine. Although both types of glands produce sweat that is odorless, each contributes differently to body odor. Apocrine sweat glands, highly concentrated in the underarms, produce sweat containing organic compounds that are decomposed by bacteria to produce body odor. Eccrine sweat glands, present all over the body but concentrated in the forehead, underarms, palms of the hands, and soles of the feet, provide the moist environment needed by the bacteria that decompose the apocrine sweat. A person's genetic makeup determines the amount and distribution of his or her sweat glands. The science behind how these glands contribute to odor will be covered in Section 3 of Chapter 4.

Personal hygiene is another major factor affecting body odor. Since natural body odors are judged in some cultures to be unpleasant, people in these cultures often try to control body odor by washing their skin and using deodorants and antiperspirants. A person's odor is also influenced by his or her oral hygiene. Brushing and flossing the teeth regularly help prevent bad breath odors.

## Section 1 Background: Your Unique Scent

Many other factors also contribute to a person's distinctive body odor. For example, one study showed that body odor is related to cultural habits such as diet. Participants in the study could detect the odors of cabbage, garlic, curry (and other spices), butter (and other fats), and fish on the skin of people who regularly consumed large quantities of these foods as part of their diet.

*Human scent control merchandise*

**Human Scent Control** ▶ You may sometimes worry about body odor, but usually a simple shower takes care of any problem. Not so for a deer hunter wanting to sneak up on a whitetail deer that can smell humans from a distance of half a mile.

How can hunters avoid being detected by deer? Although standing downwind from the deer helps, some hunters rely on special products and cleaning techniques to eliminate or mask their body odor. They use specially formulated, scent-reducing soap, shampoo, and deodorant and brush their teeth with baking soda. They wear suits containing odor-absorbing activated carbon and wash their other clothing in unscented laundry detergent that destroys odor-causing bacteria. Foot and hand odor is controlled by wearing tall rubber boots and fabric gloves. Even hunting equipment is sprayed with a scent eliminator, wiped down with a scent-free rag, and stored in large plastic bags.

# Section 1 Experiment: Scent Detectors

Is your hand scent unique?

## Procedure

### Part A: Your Own Scent

Begin this experiment by smelling your own hands. Does your left hand smell the same as your right hand? Does the hand you use most often (your dominant hand) have a stronger scent? Your dominant hand probably comes into contact with different objects in the environment more frequently, so it might have a different or stronger scent than your nondominant hand.

**Question 1:** *Write a short description of your hand scent. Use descriptive words like sweet, musky, sweaty, fruity, oily, pencil-like, cheesy, oniony, and garlicky. Do your two hands smell the same or different?*

### Part B: Identifying Gender by Scent

You will observe a class demonstration in which blindfolded volunteers, called scent detectors, attempt to identify the gender of four other volunteers based solely on the scent of their hands. So that you don't interfere with experimental results, you and other students in the class should remain quiet and not provide any clues or respond to any guesses.

Record the guesses of the scent detectors in the Part B Data Table, making notes about how easily the scent detectors seem to make each guess.

### Part C: Identifying Volunteers by Scent

You will observe a class demonstration in which blindfolded scent detectors are asked to identify the volunteers based only on hand scent. In the Part C Data Table, record the results of this challenge.

**Question 2:** *How accurate were the scent detectors' guesses in each part? Discuss possible reasons for the results.*

**Question 3:** *Was there a difference in accuracy between the male scent detectors and the female scent detectors?*

**Question 4:** *Organizers of a body odor study at the University of California Medical Center instructed study participants not to use soap, shampoo, perfume, or deodorant within 24 hours of the scent test. Explain why.*

Chapter 3: The Origins of Body Odor

## Section 1 Experiment: Scent Detectors

### Part B Data Table

| Gender of Hand Scent Volunteer | Female Scent Detector | | Male Scent Detector | |
|---|---|---|---|---|
| | Guess | Right or Wrong? | Guess | Right or Wrong? |
| | | | | |
| | Notes: | | Notes: | |
| | | | | |
| | Notes: | | Notes: | |
| | | | | |
| | Notes: | | Notes: | |
| | | | | |
| | Notes: | | Notes: | |

### Part C Data Table

| Name of Hand Scent Volunteer | Female Scent Detector | | Male Scent Detector | |
|---|---|---|---|---|
| | Guess | Right or Wrong? | Guess | Right or Wrong? |
| | | | | |
| | Notes: | | Notes: | |
| | | | | |
| | Notes: | | Notes: | |
| | | | | |
| | Notes: | | Notes: | |
| | | | | |
| | Notes: | | Notes: | |

# Instructor Notes for Section 1

Part A of this experiment demonstrates that every person has a unique scent. Students are asked to describe and compare the scent of their dominant and nondominant hands. In Parts B and C, class demonstrations determine if a blindfolded scent detector can guess the gender and identity of people based on hand scent.

## Time Required
Setup: 5 minutes
Procedure: 30 minutes
Cleanup: none

## Procedure Notes
On the day prior to the class demonstration, ask all people participating in the class demonstration portion of this experiment not to wear perfume, cologne, or other heavily scented products on the day of the demonstration.

Volunteers offering their hands to be smelled should avoid washing their hands within 15 minutes of coming to class. About 5 minutes before the demonstration begins, these volunteers should rinse their hands with water.

Be sure students understand the role of each participant in the experiment before they volunteer. Student volunteers should be briefed in advance about their role and the procedures to be followed. To avoid the issues that might be associated with smelling the hands of student volunteers, you may wish to recruit adult volunteers (for example, fellow teachers or administrators) for this role.

Another alternative procedure for Part B is to ask volunteers to wear disposable cotton gloves on the evening before the experiment. Before bedtime, have them place the worn gloves in a labeled zipper-type plastic bag. In class the next day, ask students to smell the contents of the bag and identify whether the gloves were worn by a male or a female.

### Procedure for Part B: Identifying Gender by Scent
❶ Conduct a class demonstration on identifying the gender of people by scent. Ask one male and one female to be scent detectors and two males and two females to be identified by hand scent. Remind students to be sensitive to volunteers' feelings. Seat one of the scent detectors in a chair and place a blindfold over his or her eyes.

Chapter 3: The Origins of Body Odor

**Instructor Notes for Section 1**

❷ Without revealing the name of the first volunteer, have the volunteer place his or her nondominant hand close enough to the detector's nose to be sniffed but not close enough to be touched. Ask the scent detector to sniff the volunteer's hand and guess the gender of the volunteer based on scent. Have the class record the results in the Part B Data Table.

❸ Repeat step 2 with the remaining three hand-scent volunteers.

❹ Blindfold the other scent detector and repeat steps 2 and 3 using the same hand-scent volunteers.

**Procedure for Part C: Identifying Volunteers by Scent**

❶ Continue the class demonstration or have students work in groups to see if students can identify people based on hand scent. Before putting on the blindfold, ask one male and one female scent detector to smell the nondominant hands of four volunteers. In this way the detectors learn the scents of the volunteers.

❷ Blindfold one of the scent detectors. Have the scent detector sniff a volunteer's hand and guess the identity of the volunteer based on scent memory. (The detector will know that the scent belongs to one of the four volunteers.) Ask the observers to record the results in the Part C Data Table. Repeat the process with the other three hand-scent volunteers.

❸ Repeat step 2 with the other scent detector.

## Answers and Observations

**Question 1:** *Write a short description of your hand scent. Use descriptive words like sweet, musky, sweaty, fruity, oily, pencil-like, cheesy, oniony, and garlicky. Do your two hands smell the same or different?*

Answers will vary. A typical answer might be: "Musky smell. The right hand was slightly stronger than the left."

**Question 2:** *How accurate were the scent detectors' guesses in each part? Discuss possible reasons for the results.*

Answers will vary depending on the results.

**Question 3:** *Was there a difference in accuracy between the male scent detectors and the female scent detectors?*

Answers will vary depending on the results. However, research suggests that females are more sensitive to odors than males.

**Question 4:** *Organizers of a body odor study at the University of California Medical Center instructed study participants not to use soap, shampoo, perfume, or deodorant within 24 hours of the scent test. Explain why.*

Since most personal hygiene products contain fragrance designed to cover up body odors, use of these products might mask personal scent and therefore affect the results.

## Explanation

Michael J. Russell, a research psychologist at the University of California Medical Center in San Francisco, performed a study relating odor to gender. In the study, students identified by smell alone (with about 75% overall accuracy) whether T-shirts had been worn by males or females. The T-shirts had been worn as undergarments with the wearers using no soaps, perfumes, or deodorants for 24 hours before the test. The students generally described male odor as "musky" and female odor as "sweet."

When students smell their hands in Part A, they might smell a variety of different odors. In addition to detecting their personal scent, they could smell odors from environmental sources. For example, hands pick up odors from food, coins, and wooden pencils.

The results obtained from the class demonstrations in Parts B and C greatly depend on the abilities of the scent detectors. Research suggests that females are more sensitive to odors than males. Therefore, girls may have a higher success rate in this experiment.

# Section 2 Background: Everything You Always Wanted to Know About Sweat

The body temperature of "cold-blooded" (poikilothermic) animals fluctuates with the environment. For example, a turtle basking in the sun could have an internal temperature of 80.0°F (26.7°C) and a turtle swimming in cold water could have an internal temperature of 65.0°F (18.3°C). In contrast, "warm-blooded" (homeothermic) animals must maintain a constant body temperature, regardless of the environment. For example, a teen playing basketball on a hot day has an internal temperature of about 98.6°F (37.0°C), and a teen standing in a blizzard also has an internal temperature of about 98.6°F (37.0°C). Sweating helps to cool the body when the body temperature rises above about 99.0°F (37.2°C).

The average person has about 2 million sweat glands that are either eccrine or apocrine. (See Figure 3-1.) These two types of sweat glands differ in size, location, the age that they become active, the composition of their sweat, and the conditions that trigger their activity. The smaller eccrine sweat glands are located all over the body, although they are more abundant in the forehead, underarms, palms of hands, and soles of feet. These glands are long, coiled tubes of cells that connect to pores on the skin's outer surface by long ducts. They are active from birth and produce sweat that is mostly water and salt. Eccrine glands are triggered by exercise, hot temperatures, and emotional states such as fright, nervousness, embarrassment, anger, or anxiety.

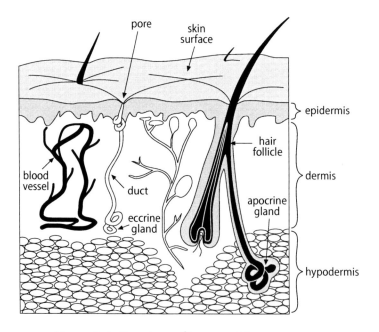

*Figure 3-1: Structure of human skin*

Apocrine sweat glands are larger than eccrine glands and are found mainly in the underarms and genitals. They are also coiled tubes but typically end in the hair follicles rather than in skin pores. Apocrine glands become active at puberty. These glands secrete a viscous (thick), sometimes milky or yellowish, odorless sweat that contains proteins, lipids (such as fatty acids and cholesterol), and steroids. Bacteria decompose or modify components of apocrine sweat to produce body odor with the characteristic "sweaty" smell in the underarms. Apocrine glands are emotionally triggered and do not play a role in temperature regulation.

Heat-induced sweating begins on the forehead, then spreads downward over the rest of the body. Emotionally-induced sweating usually occurs in the hands, feet, face, and underarms.

To deal with small increases in temperature, the body cools itself by radiating (giving off) heat into the air. Fine, thin capillary blood vessels near the surface of the skin swell and release the excess heat, resulting in the flushed look one has when feeling very warm. When this radiation method does not release enough heat, temperature-sensitive brain cells in the hypothalamus make the eccrine sweat glands active. As sweat evaporates from the surface of the skin, heat is removed from the skin and the body is cooled. In extreme conditions, the body can produce as much as 2.5 liters (about 2.7 quarts) of perspiration an hour and 10–15 liters (2.6–3.9 gallons) a day, although the average amount of sweat produced in a day is considerably less.

According to the Society of Thoracic Surgeons, one out of every 100 people sweats excessively, even when not overheated. These people perspire more from their underarms, palms, and feet than from other areas of the body. This condition, called hyperhidrosis, can be treated by a doctor.

Bathing with soap temporarily removes surface oil and sweat, which contain odor-causing bacteria. Bacteria quickly return from the environment and from populations that live in a person's oil gland ducts. Deodorants and antiperspirants help control underarm odor, where most of the non-heat-regulating (apocrine) glands are located.

# ● Section 2 Experiment: Using a Sweat Meter

Maybe a toy Fortune Teller Fish can't really tell your fortune, but it can act as a sweat meter!

## Materials
Per student
- Fortune Teller Fish in its wrapper
- paper towel
- water

## Procedure
### Part A: Discover the Secret

❶ Remove a cellophane fish from its wrapper and save the wrapper for step 2. Lay the fish on the palm of your hand and observe its behavior.

**Question 1:** *What did you observe? Discuss what factors might have made the cellophane fish act as it did.*

❷ Place the wrapper on the palm of your hand. Lay the cellophane fish on top of the wrapper and observe the behavior of the fish.

**Question 2:** *What did you observe? Based on your observations in steps 1 and 2, formulate a hypothesis on what caused the fish to behave as it did.*

❸ Slightly dampen a folded paper towel with water and squeeze out as much excess water as possible. Place the cellophane fish on the moist paper towel and observe the behavior of the fish.

**Question 3:** *Compare and contrast the fish's behavior in the above experiments. Which variable had the greatest effect on the fish's motion?*

### Part B: Design Your Own Experiment

❶ Design an experiment that uses the cellophane fish to determine if physical activity affects the amount of sweat in the palms of your hands or the rate of its evaporation from the palms. Make sure to specify the type of physical activity and its duration.

❷ Have your experiment okayed by your teacher, then try it. Record your results.

**Question 4:** *Did your physical activity affect the amount of sweat on the palm of your hand?*

# Instructor Notes for Section 2

Part A of this experiment uses a moisture-absorbing cellophane fish to demonstrate that people sweat through the palms of their hands. By comparing the fish's behavior in various settings, students deduce that moisture is a key factor for the toy fish's movement. In Part B, students design an experiment to determine whether an increase in physical activity increases sweat production in the palms of the hands.

## Time Required
Setup: none
Procedure: 20 minutes
Cleanup: none

## Materials
Per student
- Fortune Teller Fish in its wrapper

 *Fortune Teller Fish (PR9906S for 30 or PR9906L for 144) can be purchased from Terrific Science Books, Kits, & More; 866/438-6724; http://www.tsbkm.com.*
- paper towel
- water

## Answers and Observations

**Question 1:** *What did you observe? Discuss what factors might have made the cellophane fish act as it did.*

When the fish is placed on a bare palm, the fish should begin to curl. Heat and/or moisture from the hand are the most likely factors in causing the fish to curl. Other common hypotheses are wind, air currents, humidity, and static charge.

**Question 2:** *What did you observe? Based on your observations in steps 1 and 2, formulate a hypothesis on what caused the fish to behave as it did.*

The fish does not curl when placed on the plastic wrapper. Since the wrapper acts as a barrier from the hand's moisture but not from the hand's heat, the hand's natural moisture (sweat) might be a key factor in the fish's motion.

**Instructor Notes for Section 2**

**Question 3:** *Compare and contrast the fish's behavior in the above experiments. Which variable had the greatest effect on the fish's motion?*

When placed directly on the palm of the hand, the fish moves more like when it is placed on the moist paper towel than when placed on the wrapper on the palm of the hand. Moisture appears to have the greatest effect on the fish's motion.

**Question 4:** *Did your physical activity affect the amount of sweat on the palm of your hand?*

Most people will experience an increase in palm sweating with an increase in physical activity, although results will vary depending on the intensity of the physical activity and the humidity in the air. If body temperature increases during the physical activity, the rate of sweat evaporation will also increase.

## Explanation

The Fortune Teller Fish is made of a special type of cellophane that absorbs moisture. (The cellophane available in craft or floral shops does not absorb sufficient amounts of moisture to be effective in this experiment.) Fortune Teller Fish curl and bend when placed in the palm of a hand because the cellophane absorbs the sweat from the hand, heat from the hand warms this moisture, and the heat causes the moisture to evaporate and the ends of the fish to curl. The fish is less apt to move if a plastic barrier like the wrapper is placed between the hand and the fish because the plastic does not allow significant amounts of moisture to pass through it.

**Instructor Notes for Section 2**

*Overhead Master: Structure of Human Skin*

**Chapter 3: The Origins of Body Odor**

# Section 3 Background: No Sweat

Mammals, including people, are warm-blooded (homeothermic) animals that often use sweating as a mechanism to prevent overheating. How does sweating cool the body? Your body is cooled as water from your sweat evaporates. In the evaporation process, liquid water molecules absorb heat energy from the skin and enter the gas state. The gaseous water molecules carry this energy away, leaving the skin feeling cooler.

When all of a person's eccrine sweat glands are working at their maximum, the rate of perspiration can reach 2.5 liters (about 2.7 quarts) per hour. This can result in dangerous fluid loss. For this reason, people playing sports must be sure to drink fluids at regular intervals.

What if you couldn't sweat at all? People who are quadriplegic have paralyzed arms and legs. Quadriplegics are often unable to sweat. Therefore, a quadriplegic may have to stay in temperature-controlled environments, especially during hot weather, to keep from overheating.

People can also reduce their body temperature by applying an external source of moisture to the skin and allowing the moisture to evaporate. Runners, bicyclists, and others who spend considerable time outdoors sometimes use cloth collars with polymer crystals sewn inside. These crystals are capable of absorbing hundreds of times their own weight in water. The collar is soaked in water for several hours and worn around the neck. The evaporation from the collar helps cool the body the same way that sweat evaporation does. The advantage of using a cool collar over a plain wet cloth is that the crystals release their water very slowly, affording longer periods of cooling before the collar has to be wetted again.

Although most mammals have sweat glands that aid in regulating temperature, not all mammals do. For example, the lack of sweat glands in kangaroo rats is one of a set of adaptations to help their bodies conserve water in extreme desert heat. To avoid overheating, kangaroo rats spend the heat of the day in their moist, humid burrows and only come out at night when the air is cool. Elephants only have sweat glands near their toenails. To keep cool, they use their trunks to cover themselves with water, dry dust, or mud. They also flap their large ears to create a breeze and cool the blood within the ears' large blood vessels.

# Section 3 Experiment: Sweating Is Cool

How does sweating cool us off?

## Materials
Per group
- water in a cup or container
- test liquid in a dropper bottle
- laboratory thermometer that reads from 0–30°C
- lump of clay
- cotton ball

## Safety and Disposal
Some of the liquids used in this experiment are flammable, and their vapors are irritating to the eyes and respiratory system. Use these liquids only in a well-ventilated area and keep flames away. The liquids are harmful if ingested and can cause severe damage to the eyes. Goggles must be worn during this experiment. Should contact with the eyes occur, rinse with water for 15 minutes and seek immediate medical attention.

Liquids can be saved for future use or flushed down the drain with water.

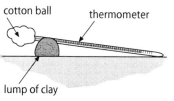

*Figure 3-2: Setup for experiment*

## Procedure

1. Dip your finger in water and then hold your finger in the air.

   **Question 1:** *Describe what you feel. Why does your skin feel this way?*

2. Wrap a cotton ball around the bulb end of a thermometer.

3. Prop the thermometer on a lump of clay so that the cotton does not touch the desktop. (See Figure 3-2.)

4. Record the liquid you are testing. Read and record the temperature of the thermometer.

5. Place 10 drops of the test liquid onto the cotton ball.

6. Record the temperature every 2 minutes for 16 minutes.

7. Calculate the difference between the starting temperature and the lowest temperature reached. This is the overall temperature change for your test liquid. Share this information with the class. As a class, complete the Class Data Table and construct a bar graph of overall temperature changes for the test liquids.

**Chapter 3: The Origins of Body Odor**

## Section 3 Experiment: Sweating Is Cool

| Class Data Table for Temperature Changes ||
|---|---|
| Test Liquid | Overall Temperature Change |
|  |  |
|  |  |
|  |  |
|  |  |
|  |  |
|  |  |
|  |  |
|  |  |
|  |  |
|  |  |

**Question 2:** *Why did the temperature of the thermometer bulb drop as the liquid evaporated from the cotton ball surrounding the bulb?*

**Question 3:** *Of all the test liquids used in the class, which would you choose to cool your body most effectively in a short amount of time? Are safety issues involved in your choice?*

**Question 4:** *How does the evaporation of sweat from skin help to regulate body temperature?*

# Instructor Notes for Section 3

In this experiment, students dip their fingers in water to feel the cooling effect of evaporation. Students measure the change in temperature as various test liquids evaporate and then graph their results. Students relate their findings to how sweating cools the body and regulates body temperature.

## Time Required
Setup: 10 minutes
Procedure: 25–30 minutes
Cleanup: 10 minutes

## Materials
For Getting Ready
- dropper bottle for each test liquid
- masking tape and marker for labels
- some of the following test liquids:
  - water
  - nail polish remover containing either acetone or methyl ethyl ketone
  - rubbing alcohol (70% isopropyl alcohol solution)
  - perfume
  - cologne
  - aftershave lotion
  - preshave lotion
  - oil
  - artificial or natural vanilla or other flavor extract
  - mouthwash

For the Procedure
Per group
- water in a cup or container
- test liquid in a dropper bottle
- laboratory thermometer that reads from 0–30°C
- lump of clay
- cotton ball

## Safety and Disposal

Some of the liquids used in this experiment are flammable, and their vapors are irritating to the eyes and respiratory system. Use these liquids only in a well-ventilated area and keep flames away. The liquids are harmful if ingested and can cause severe damage to the eyes. Goggles must be worn during this experiment. Should contact with the eyes occur, rinse with water for 15 minutes and seek immediate medical attention.

Certain liquids are inappropriate for use in the experiment. For example, methanol and duplicating fluid are toxic when absorbed through the skin or when the fumes are inhaled. For this reason, use only the liquids suggested in the Materials list.

Liquids can be saved for future use or flushed down the drain with water.

## Getting Ready

Fill dropper bottles with test liquids and label the bottles appropriately.

## Sample Data

| Typical Temperature Changes for Some Liquids | | | |
|---|---|---|---|
| Liquid | Room Temperature | Lowest Temperature | Temperature Change |
| vanilla extract | 25°C | 20°C | −5°C |
| banana extract | 25°C | 21°C | −4°C |
| nail polish remover (containing acetone) | 25°C | 12°C | −13°C |
| water | 25°C | 19°C | −6°C |
| rubbing alcohol | 25°C | 17°C | −8°C |

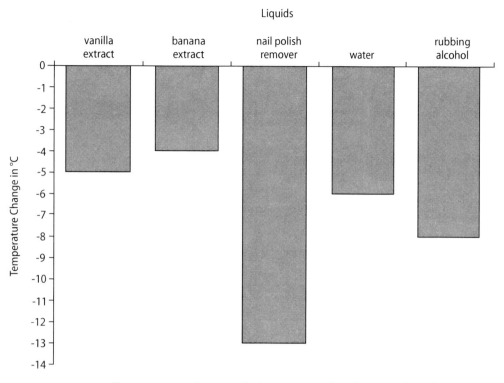

*Temperature changes during evaporation for some liquids*

## Answers and Observations

**Question 1:** *Describe what you feel. Why does your skin feel this way?*

Skin moistened with water will feel cooler because of the cooling effect of evaporation.

**Question 2:** *Why did the temperature of the thermometer bulb drop as the liquid evaporated from the cotton ball surrounding the bulb?*

When liquids evaporate, they absorb heat energy from the surroundings. This causes the temperature of the surroundings to be lowered.

**Question 3:** *Of all the test liquids used in the class, which would you choose to cool your body most effectively in a short amount of time? Are safety issues involved in your choice?*

To answer this question, students would select the test liquid with the greatest temperature change from the bar graph. Students would also have to consider possible hazards from these chemicals contacting the skin. They should read product labels and do web research to determine the safety considerations of their choice.

**Instructor Notes for Section 3**

**Question 4:** *How does the evaporation of sweat from skin help to regulate body temperature?*

Sweat is composed mainly of water. When sweat evaporates from the skin, the water absorbs some of the heat energy from our bodies to change from a liquid to a gas. This evaporation process lowers the temperature of our bodies.

## Explanation

The evaporation of a liquid requires energy as the liquid changes to a gas. As a liquid evaporates, it removes heat energy from its surroundings. This removal of energy is reflected in the temperature drop felt on the skin and seen on the thermometer in this experiment. As a test liquid evaporates from the cotton ball, energy is absorbed by the liquid and lost by the surroundings. This causes the thermometer's temperature to fall below the initial room temperature reading.

*Overhead Master*

| Class Data Table for Temperature Changes ||
|---|---|
| Test Liquid | Overall Temperature Change |
|  |  |
|  |  |
|  |  |
|  |  |
|  |  |
|  |  |
|  |  |
|  |  |
|  |  |
|  |  |

# Chapter 4
# When Life Stinks

Chapter 4 Overview ........................................................................................................... 72
National Science Education Standards ....................................................................... 72
Cross-Curricular Integration............................................................................................ 73

Section 1 Background: Sweaty Feet............................................................................. 75
Section 1 Experiment: Ban the Rotten Sneaker ........................................................ 77
Instructor Notes for Section 1 ........................................................................................ 80

Section 2 Background: Beware the Fire-Breathing Dragon!................................. 84
Section 2 Experiment: Spray It Away............................................................................ 86
Instructor Notes for Section 2 ........................................................................................ 89

Section 3 Background: Control and Conquer .......................................................... 92
Section 3 Experiment: Life in the Pits........................................................................... 94
Instructor Notes for Section 3 ........................................................................................ 96

# Chapter 4 Overview

**Key Science Topics**
- parts of the body prone to odor
- origins of foot and breath odor
- ability of substances to absorb, mask, or neutralize odors

It's a fact of life that the physical changes of puberty include an increase of hair oil, foot odor, and underarm sweating and odor. These changes can cause teenagers to become self-conscious and worried about social acceptance. Imagine the embarrassment of changing into dance or athletic shoes and having your friends back away because of the smell of your feet, or raising your hand in class to reveal a large wet spot under your arm. These are real worries that teenagers deal with every day.

This chapter presents the chemistry and physiology of body odors and how to control these odors. Section 1 describes the causes of foot odor and ways to address the problem. In the Section 1 experiment, students investigate how shoe inserts work to control odors. Section 2 focuses on mouth odor. In the Section 2 experiment, students investigate the effectiveness of breath sprays as a temporary means of controlling unpleasant-smelling breath. Section 3 addresses underarm odor. A student experiment simulates the growth of bacteria under the arm and involves students in testing the effectiveness of several substances to control that growth.

## National Science Education Standards
This chapter addresses the National Science Education Content Standards for grades 5–8 and 9–12 as described in the following lists.

### Section 1
*Science as Inquiry:*
Abilities Necessary to Do Scientific Inquiry
- Students design and conduct an investigation on the effectiveness of shoe inserts in controlling the odor of butyric acid and similar compounds.
- Students develop an explanation for how various shoe inserts control foot odor based on evidence gained through this experiment.
- Use appropriate tools and techniques to gather, analyze, and interpret data. Students calculate weighted class averages of their results.

*Physical Science:*
Properties and Changes of Properties in Matter
- Substances react chemically in characteristic ways with other substances to form new substances (compounds) with different characteristic properties. Students explore how odor is controlled using baking soda (acid/base

reaction which neutralizes butyric acid to form water and a salt) and activated charcoal (which bonds odor molecules).

*Science and Technology:*
Abilities of Technological Design
- Students evaluate the effectiveness of various brands of odor-control shoe inserts in controlling test odors.

## Section 2
*Science as Inquiry:*
Abilities Necessary to Do Scientific Inquiry
- Students conduct an investigation to determine the effectiveness of various breath sprays at masking odorous test substances.

*Science and Technology:*
Abilities of Technological Design
- Students evaluate the effectiveness of various brands of breath spray in masking odorous test substances.

## Section 3
*Science as Inquiry:*
Abilities Necessary to Do Scientific Inquiry
- Students conduct an investigation into the factors that affect yeast growth.

*Life Science:*
Structure and Function of Living Systems
- Cells carry on the many functions needed to sustain life. They grow and divide, thereby producing more cells. This requires that they take in nutrients, which they use to provide energy for the work that cells do and to make the materials that a cell needs. Students investigate the factors that affect the growth of yeast.

## Cross-Curricular Integration
The experiments in this chapter can be integrated with other areas of the curriculum to emphasize the relationships between subjects. Some ideas for cross-curricular integration are listed below.

## Food Science
- Have students research the role of bacteria in the production of foods such as yogurt, buttermilk, and cheese.
- Have students research the role of bacteria in food spoilage.

## Chapter 4 Overview

### Health
- Have students investigate how sweaty feet can lead to fungal problems and what precautions can be taken to avoid these problems.
- Relate the section on bad breath to health issues involving smoking.

### Life Science
- Have students investigate the role of bacteria in the decomposition of matter.

### Physical Science
- Integrate this unit with the study of chemical reactions in the body that produce gases.

### Social Science
- Have students research attitudes in various cultures towards body odor. For example, in some cultures the word for *kissing* is the same as the word for *smelling*.
- Have students survey advertising claims for various foot and underarm deodorants and breath sprays. Based on what they've learned in this unit, have them evaluate these claims.

# ● Section 1 Background: Sweaty Feet

According to the Society of Chiropodists and Podiatrists, we have more sweat glands per square inch in our feet than anywhere else in our bodies. Sweat glands in your feet secrete sweat all the time to keep the skin moist. However, foot sweating can increase due to a warm environment, physical stress on the feet (from exercise or excessive standing), emotional stress, or an inherited condition.

Adolescents tend to have sweaty feet because changing hormonal levels in the body can cause overactive sweat glands. Have you noticed that your feet seem to sweat more when you're wearing socks or closed shoes made from synthetic materials? This is because many synthetic materials don't "breathe"; that is, they don't allow the passage of air or vapor, so the foot moisture is trapped.

Bacteria thrive in the warm, moist environment inside your shoes. The constant friction of walking rubs dead skin cells off your feet, and these cells feed the bacteria. The waste produced by these bacteria contains butyric acid, a molecule with an unpleasant, rancid odor. Unappetizing as it may seem, butyric acid is also produced when bacteria feed on milk proteins and milkfat to make cheese! (This explains the resemblance between foot odor and some cheese odors.)

**Zits** / by Jim Borgman and Jerry Scott

You can reduce foot odor by periodically going barefoot, wearing cotton socks, or wearing shoes made from materials that breathe. Thick cotton socks can absorb some of the sweat, wicking it away from the odor-causing bacteria. Shoes made of cotton canvas, natural leather, or with open construction permit evaporation of moisture. Allow shoes to air out for at least 24 hours

### Section 1 Background: Sweaty Feet

before wearing them again. This allows the moisture and bacterial waste buildup inside the shoe to dry up, which helps reduce foot odor. If all else fails, a severely sweaty foot problem can be treated with rubbing alcohol, tea foot baths, antiperspirants, or medication.

*How much do your sneakers stink?*

**The Stinkier, the Better** ▶  Do you think your sneakers are stinky enough to win a contest? Well, kids of ages 5–15 can enter their smelly, worn-out sneakers in the annual Odor-Eaters® Rotten Sneaker Contest™. Competitors from all over the United States begin by participating in local or online contests to win a year's supply of Odor-Eaters and an all-expense-paid trip to the national contest in Montpelier, VT.

During the national contest, judges smell and rate sneakers based on offensive odor, worn-out soles, frayed laces, and torn toes. Contestants must also tell the judges how their sneakers got so rotten. The grand-prize winner receives a $500 U.S. Savings Bond, a year's supply of Odor-Eaters, and the chance to have his or her sneakers enshrined in the Odor-Eaters Hall of Fumes. Check out the details at *http://www.odor-eaters.com*.

# ● Section 1 Experiment: Ban the Rotten Sneaker

How do odor-control shoe inserts work?

## Materials

Part A
Per group
- 4 zipper-type freezer bags
- test substance
- knife (if the test substance is solid and needs chopping)
- newspaper (if the test substance is solid and needs chopping)
- 2.5-mL (½-teaspoon) measuring spoon (if the test substance is solid)
- 4 cotton balls (if the test substance is liquid)
- eyedropper (if the test substance is liquid)
- permanent marker
- 15-mL (1-tablespoon) measuring spoon
- baking soda
- activated charcoal
- piece of odor-control shoe insert
- scissors

Part B
Per group
- piece of each brand of odor-control shoe insert
- scissors
- hand lens or microscope

Part C
Per student
- different types of cheese
- other materials needed for your experiment

## Safety and Disposal

Be careful when using the knife to chop test substances. As with all food items used in a laboratory setting, do not eat any of the food used in this experiment.

Chapter 4: When Life Stinks

Section 1 Experiment: Ban the Rotten Sneaker

## Procedure
### Part A: Setting Up the Experiment (Day 1)

❶ Working in groups, prepare four freezer bags with the assigned test substance.

   a. If your test substance is solid and chopping is needed, carefully chop any big pieces on some newspaper. Place about 2.5 mL (½ teaspoon) of the chopped test substance in the bottom of each bag. Label each bag with your group's name.

   b. If your test substance is liquid, prepare four cotton balls by putting 8–10 drops of the liquid on each. Make sure the cotton is not dripping wet. Place a cotton ball in the bottom of each bag. Label each bag with your group's name.

❷ Continue preparing the four bags as described below.

   a. Add 45 mL (3 tablespoons) baking soda to one of the bags and label appropriately.

   b. Add 45 mL (3 tablespoons) activated charcoal to another bag and label appropriately.

   c. Cut your piece of a shoe insert into very small pieces using scissors. Add 45 mL (3 tablespoons) tightly packed insert pieces to a third bag, and label appropriately.

   d. Label a fourth bag containing just the test substance with the word "control."

   e. Seal the bags tightly and shake thoroughly to mix the ingredients.

❸ Store the bags at room temperature until Day 2 of the experiment. Make sure the bags with different test substances are not stored in the same location, since the odor from one bag could contaminate other bags.

### Part B: Assessing and Recording the Results (Day 2)

❶ Each group will set up a station with their four bags. Students will visit each station.

❷ At each station, open the control bag and smell its contents by using the wafting technique. One at a time, open and smell each of the other bags, and rank the intensity of the test substance odor from each bag as compared to the odor of the control bag. Rate the odor in each bag as 1 Very Strong, 2 Strong, 3 Moderate, 4 Weak, or 5 Very Weak. Record your ratings. To empty your olfactory receptors, sniff your sleeve after you sniff each bag.

**Section 1 Experiment: Ban the Rotten Sneaker**

❸ Share your results with the class and complete a class data table for each substance.

**Question 1:** *Which odors were most completely absorbed by the a) baking soda, b) activated charcoal, and c) shoe insert you tested?*

❹ Obtain samples of several different brands of shoe inserts from your teacher. Read the labels and make a list of ingredients in each brand. Dissect a piece of each brand and use a hand lens or microscope to observe its physical structure. Draw and label a picture of what you find for each sample.

**Question 2:** *Explain how the brand of shoe insert you tested can control foot odor.*

### Part C: Design Your Own Experiment

❶ Design an experiment using smelly cheese to determine how effective odor-control shoe inserts are in controlling the odor of butyric acid and similar compounds.

❷ Have your experiment okayed by your teacher. Try your experiment and record your results.

❸ Compare your results with those of the rest of the class.

**Question 3:** *How was your experiment similar to those designed by the rest of the class and how was it different?*

**Question 4:** *Were your results the same or different than the results obtained by most of the class? Why?*

# Instructor Notes for Section 1

This experiment explores the use of shoe inserts in controlling foot odor. Students investigate the effectiveness of baking soda, activated charcoal, and odor-control shoe inserts in controlling various test substance odors. Students then compare the ingredients and physical structures of different brands of odor-control shoe inserts. They design their own experiments to determine how effective shoe inserts are in controlling the odor of butyric acid and similar compounds.

## Time Required

Setup: 10 minutes
Procedure: Part A, 15 minutes
　　　　　Part B, 20 minutes
　　　　　Part C, time will vary
Cleanup: 10 minutes

## Materials

For Getting Ready
- newspaper
- hammer
- scissors
- several packages of different brands of odor-control shoe inserts (such as Odor-Eaters® and Dr. Scholl's®)

For Part A
Per group
- 4 zipper-type freezer bags
- 1 of the following test substances:
  - onion
  - garlic
  - table salt
  - powdered spice
  - orange rind
  - grapefruit rind
  - liquid natural flavoring
  - liquid artificial flavoring
  - liquid smoke
  - rubbing alcohol

- knife (if the test substance is solid and needs chopping)
- newspaper (if the test substance is solid and needs chopping)
- 2.5-mL (½-teaspoon) measuring spoon (if the test substance is solid)
- 4 cotton balls (if the test substance is liquid)
- eyedropper (if the test substance is liquid)
- permanent marker
- 15-mL (1-tablespoon) measuring spoon
- baking soda
- activated charcoal

*Purchase activated charcoal commonly used for aquariums.*
- about half of an odor-control shoe insert

*Have at least two groups use each test substance but different brands of shoe insert.*
- scissors

For Part B
Per group
- piece of each brand of odor-control shoe insert
- scissors
- hand lens or microscope

For Part C
Per student
- different types of cheese, including some with strong odors, such as Limburger or Parmesan
- other materials needed for the student-designed experiment

## Safety and Disposal

Caution students to be careful when using the knife to chop test substances. Caution students never to eat food that is used in the laboratory.

## Getting Ready

Place the activated charcoal between layers of newspaper and hit with a hammer to break up large chunks. Cut different brands of odor-control shoe inserts so each group will have about half an insert of one brand for Part A and a piece of each brand for Part B.

## Procedure Notes

To introduce the experiment, ask students why sneakers are smelly. (As bacteria feed on perspiration, they create waste, which has an odor that is

**Instructor Notes for Section 1**

transferred to shoes.) Ask students how foot odor can be controlled. (Reduce the cause of the foot odor by reducing the amount of perspiration or the number of bacteria; or trap or neutralize the odor after it is produced.)

On Day 2, set up stations around the classroom so that each station contains all of the bags from one group. After all students have visited each station and recorded their results, collect class data by compiling the results of each test substance experiment on a separate class data table. Calculate weighted class averages and discuss the results.

## Sample Data

| Sample Data Table for Shoe Insert Test | | | | | |
|---|---|---|---|---|---|
| Test Substance | Control | Baking Soda | Activated Charcoal | Dr. Scholl's | Odor-Eaters |
| orange | weak odor | no odor | no odor | weak odor | weak odor |
| garlic | moderate odor | weak odor | no odor | moderate odor | weak odor |

## Answers and Observations

**Question 1:** *Which odors were most completely absorbed by the a) baking soda, b) activated charcoal, and c) shoe insert you tested?*

Answers will vary based on substance and brand of shoe insert tested.

**Question 2:** *Explain how the brand of shoe insert you tested can control foot odor.*

Answers will vary depending on the brand of shoe insert tested.

**Question 3:** *How was your experiment similar to those designed by the rest of the class and how was it different?*

Answers will vary depending on the individual experimental designs.

**Question 4:** *Were your results the same or different than the results obtained by most of the class? Why?*

Answers will vary depending on the results of the class.

## Explanation

Shoe inserts designed to fight foot odors contain baking soda or activated charcoal or a mixture of both. These ingredients act differently on the molecules that cause the odors. Baking soda is a base. Bases neutralize acids to form salts and water. The salts formed in this process do not have the same unpleasant odor as the starting butyric acid. Activated charcoal works differently. It contains millions of tiny pores between its carbon atoms that serve as binding sites for certain chemicals. The activated charcoal absorbs a large number of odor-causing molecules. Thus, the molecules are trapped by the charcoal and are prevented from reaching your nose.

# Section 2 Background: Beware the Fire-Breathing Dragon!

Halitosis, another name for unpleasant mouth odor or "bad breath," is not a new problem. Many ancient civilizations made tooth-cleaning powders and tools from various substances. For example, Egyptians mixed ashes of oxen hooves and eggshells with myrrh and pumice. Greeks used their fingers to rub pumice, talc, ground alabaster, coral powder, and iron rust on their teeth. Romans used gold and silver toothpicks. The Chinese invented the first known toothbrush by attaching the bristles from a horse's mane to an ivory handle.

Bad breath typically originates in the mouth where hundreds of bacterial species live and feed on food particles and sloughed-off mouth cells. These bacteria produce foul-smelling substances such as hydrogen sulfide (smells like rotten eggs), methyl mercaptan and skatole (both present in feces), cadaverine (associated with rotting flesh), putrescine (found in decaying meat), and isovaleric acid (smells like sweaty feet).

The very back of the tongue is the main site for bad breath development. This region provides an ideal site for the growth of anaerobic bacteria (those capable of living without oxygen). This area is not well cleansed by saliva, so once the bacteria start to grow, they tend to stay and multiply. Additionally, mucus dripping down the back of the throat (called post-nasal drip) gets deposited in this region and provides a source of food for the bacteria.

Americans spend billions of dollars each year on toothpaste, toothbrushes, floss, mouthwashes, mints, and other breath fresheners. Dental professionals recommend brushing your teeth as soon as possible after eating and flossing on a regular basis to prevent food from decaying in your

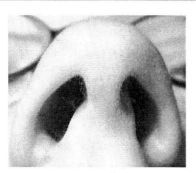

*A source of odor?*

**Mysterious Bad Breath** ▶ A 28-year-old woman went to her doctor complaining of bad breath. Although other people had noticed only a slight odor, the patient perceived a fierce odor. The doctor detected a slight odor from her mouth but also noticed a peculiar odor emanating from the patient's nose. An examination revealed that she had a foreign object up her nose, which was removed under anesthesia in a hospital. The foreign object turned out to be a child's plastic bead that was probably put in the nose about 25 years earlier!

Children sometimes put objects up their nose because they are curious, bored, or copying other children. These objects could be things such as peanuts, raisins, seeds, clay, pebbles, or pencil erasers. If the object absorbs water, it quickly becomes a good environment for bacteria to grow. These bacteria produce smelly waste products, sometimes causing a foul-smelling discharge from the nose. Objects unable to absorb water, such as plastic beads and pebbles, could theoretically stay in the nose for many years without detection. In time, these objects may become coated with calcified material, lead to low-grade infections, and eventually produce an odor.

Case study from Bad Breath Research website:
http://www.tau.ac.il/~melros/faq/33.html
by Mel Rosenberg.

mouth. At times, a tongue scraper is recommended to reduce odor-causing bacterial growth on the tongue. Some mouthwashes and breath sprays can also help freshen breath by killing bacteria, but most breath products last only about 20 minutes to 2 hours.

Bad breath can also come from odor sources other than directly from the mouth. Some foods, like onions and cabbage, contain high amounts of sulfur compounds. When these sulfur compounds are digested, they can be absorbed in the blood stream, carried by the blood to the lungs, and exhaled as bad breath. Inflamed sinuses, putrid tonsils, and hundreds of other diseases and conditions can cause bad breath. If you believe you have a bad breath problem, be sure to check with your dentist or doctor for information, particularly if the problem lasts for a extended period of time.

*A Halimeter detects bad breath.*

**Odor Meters** ▶ Do you have bad breath? How can you tell? Would you breathe into your hand and take a whiff? Try it. Good breath or bad, this technique doesn't work. Why? Because of a condition called olfactory fatigue, which makes it difficult to recognize your own breath problems. So what can you do to find out if you have bad breath? The best nonmedical way is to ask a family member or friend.

Some dentists and doctors measure bad breath with an instrument called a Halimeter®, which detects smelly sulfur compounds, including hydrogen sulfide, methyl mercaptan, and others. A tube placed in the back of the mouth takes a breath sample and, within seconds, the Halimeter detects amounts of these sulfur compounds to the parts per billion (ppb). A measurement of 120 ppb or higher indicates problem breath.

The Halimeter only measures sulfur-containing compounds, but other chemicals, such as putrescine and cadaverine, can also cause bad breath. Therefore, dentists and doctors using the Halimeter also use the organoleptic method to assess breath. This method involves smelling the patient's breath to determine odor intensity and to detect specific odors that might provide clues about medical conditions.

In testing oral hygiene products, trained odor judges use the organoleptic method to rate the intensity and the quality (neutral, pleasant, or bad) of breath odor. After the breath (usually "morning breath") of test subjects is smelled and analyzed, subjects use a test product (such as toothpaste, mouthwash, or breath spray), and their breath is tested again to assess how effective the product was in reducing bad breath.

# Section 2 Experiment: Spray It Away

Can breath spray really cover up mouth odors?

## Materials

Per group
- permanent marker
- 6 zipper-type plastic bags (pint or sandwich size)
- knife
- newspaper
- cotton ball
- 1.25-mL (¼-teaspoon) and 2.5-mL (½-teaspoon) measuring spoons
- all of the following test substances:
  - fresh onion
  - fresh garlic
  - orange, lemon, lime, or grapefruit rind
  - blue cheese or other smelly cheese
  - canned or dried fish
  - liquid smoke
- vial of breath spray

## Safety and Disposal

Be careful when using the knife to chop test substances. As with all food items used in a laboratory setting, do not eat any of the food used in this experiment. Take care to aim the breath sprays directly into the bags and to keep the spray out of your eyes.

## Procedure

❶ Working in groups, carefully chop the solid test substances into small pieces on some newspaper. Place about 2.5 mL (½ teaspoon) of each test substance in a separate plastic bag, and label the bag appropriately. Put 1.25 mL (¼ teaspoon) liquid smoke on a cotton ball, place the cotton ball in another plastic bag, and label it appropriately. Record the brand of breath spray that you've been assigned to test in the Data Table.

❷ Open each bag in turn, sniff, and immediately zip the bag shut again. Record your observations of the strength of the odor in the Data Table.

❸ Open each bag and pump two sprays of the breath spray directly into the bag. Zip the bag shut immediately after you spray.

❹ Immediately repeat step 2.

Section 2 Experiment: Spray It Away

❺ Let the bags stand for 15 minutes, then repeat step 2.

❻ Share and compare your group's results with the class.

| Data Table for Breath Spray Test | | | | | |
|---|---|---|---|---|---|
| Brand of Breath Spray Tested: | | | | | |
| Test Substance | Ultra Strong Odor | Strong Odor | Medium Odor | Weak Odor | Ultra Weak Odor |
| onion without breath spray | | | | | |
| onion with breath spray (immediate) | | | | | |
| onion with breath spray (after 15 minutes) | | | | | |
| garlic without breath spray | | | | | |
| garlic with breath spray (immediate) | | | | | |
| garlic with breath spray (after 15 minutes) | | | | | |
| citrus without breath spray | | | | | |
| citrus with breath spray (immediate) | | | | | |
| citrus with breath spray (after 15 minutes) | | | | | |
| cheese without breath spray | | | | | |
| cheese with breath spray (immediate) | | | | | |
| cheese with breath spray (after 15 minutes) | | | | | |
| fish without breath spray | | | | | |
| fish with breath spray (immediate) | | | | | |
| fish with breath spray (after 15 minutes) | | | | | |
| smoke without breath spray | | | | | |
| smoke with breath spray (immediate) | | | | | |
| smoke with breath spray (after 15 minutes) | | | | | |

## Section 2 Experiment: Spray It Away

**Question 1:** *Which original odors were most effectively masked by your brand of breath spray?*

**Question 2:** *Which original odors were least effectively masked by your brand of breath spray?*

**Question 3:** *Based on the class data, which breath spray was the most effective in masking the most odors?*

**Question 4:** *Were there any odors that were not well masked by any brand?*

**Question 5:** *Which brand provided the most long-lasting masking of the original odors?*

# Instructor Notes for Section 2

Students expose odorous test substances to different brands of breath spray to determine the sprays' effectiveness in masking the odors. Students assess which breath spray was most effective, least effective, and most long-lasting and which test substance odors were most persistent.

## Time Required

Setup: 5 minutes
Procedure: 35 minutes (15 minutes of this time is "wait time," for which other activities can be planned)
Cleanup: 10 minutes

## Materials

Per group
- permanent marker
- 6 zipper-type plastic bags (pint or sandwich size)
- knife
- newspaper
- cotton ball
- 1.25-mL (¼-teaspoon) and 2.5-mL (½-teaspoon) measuring spoons
- all of the following test substances:
  - fresh onion
  - fresh garlic
  - orange, lemon, lime, or grapefruit rind
  - blue cheese or other smelly cheese
  - canned or dried fish
  - liquid smoke
- vial of breath spray (Provide groups with different brands.)

## Safety and Disposal

Caution students to be careful when using the knife to chop test substances. Instruct students to aim breath sprays directly into the bags. You may wish to have students wear goggles to keep spray out of their eyes. Caution students never to eat food that is used in the laboratory.

Chapter 4: When Life Stinks

# Instructor Notes for Section 2

## Sample Data

| Sample Data Table for Breath Spray Test | | | | | |
|---|---|---|---|---|---|
| Brands of Breath Spray Tested: Binaca® Concentrated (B) and Binaca® Fast Blast® (FB) | | | | | |
| Test Substance | Ultra Strong Odor | Strong Odor | Medium Odor | Weak Odor | Ultra Weak Odor |
| onion without breath spray | | X | | | |
| onion with breath spray (immediate) | | B and FB | | | |
| onion with breath spray (after 15 minutes) | | | B and FB | | |
| garlic without breath spray | | X | | | |
| garlic with breath spray (immediate) | | | B and FB | | |
| garlic with breath spray (after 15 minutes) | | | B | FB | |
| citrus without breath spray | | | X | | |
| citrus with breath spray (immediate) | | | B and FB | | |
| citrus with breath spray (after 15 minutes) | | | | B | FB |
| cheese without breath spray | | X | | | |
| cheese with breath spray (immediate) | | B | FB | | |
| cheese with breath spray (after 15 minutes) | | | B | | FB |
| smoke without breath spray | | X | | | |
| smoke with breath spray (immediate) | | B and FB | | | |
| smoke with breath spray (after 15 minutes) | | B and FB | | | |

## Answers and Observations

**Question 1:** *Which original odors were most effectively masked by your brand of breath spray?*

Results will vary with the type of breath spray used. Many breath sprays contain minty odors that are remarkably successful in masking odors like garlic and onion.

**Question 2:** *Which original odors were least effectively masked by your brand of breath spray?*

Results will vary with the type of breath spray used; however, liquid smoke is a very penetrating odor and is generally more difficult to mask.

**Question 3:** *Based on the class data, which breath spray was the most effective in masking the most odors?*

Results will vary with the types of breath spray used.

**Question 4:** *Were there any odors that were not well masked by any brand?*

Results will vary with the types of breath spray used; however, liquid smoke is a very penetrating odor and is generally more difficult to mask.

**Question 5:** *Which brand provided the most long-lasting masking of the original odors?*

Results will vary with the types of breath spray used.

## Explanation

Breath sprays work to fight bad breath in several different ways, depending on their ingredients. Many breath sprays contain natural mint flavors, such as peppermint, spearmint, and wintergreen, that mask bad breath. The stronger, more pleasant odor of the mint overpowers unpleasant breath odor by saturating the nasal receptors and preventing the brain from perceiving the bad odor.

Some breath sprays also contain alcohol, which works to kill bacteria in the mouth. Oral bacteria produce waste containing volatile sulfur compounds (VSCs). Since these VSCs give breath its bad odor, reducing the oral bacteria population reduces bad breath.

A relatively new type of breath spray contains reactive oxygen atoms that neutralize the odor-producing VSCs. These sprays use an oxidizing agent, such as chlorine dioxide, to change the VSCs from active sulfide to inactive sulfate, which is odorless and tasteless. The chlorine dioxide also acts as an antimicrobial agent.

# ● Section 3 Background: Control and Conquer

When we think of body odor, we often think specifically of underarm odor. Did you know that underarm sweat has no odor? However, when sweat gets trapped under the arm and can't evaporate, bacteria feed on that sweat and produce odor.

The natural odor of an individual's underarm is unique since the exact composition and pH of sweat (which ranges from about 4 to 6) vary from person to person, depending on gender, heredity, and diet. This person-to-person variation in sweat causes environmental variation in the underarm and therefore variation in the kinds of bacteria that thrive there. Different underarm bacteria feeding on sweat give off several dozen different waste products, including compounds such as butyric acid, short-chain fatty acids, and amines (nitrogen-containing compounds). *Staphylococcus* bacteria feeding off sweat produce an acidic odor, and diphtheroid bacteria produce a more pungent odor. The compound largely responsible for body odor is a short-chain fatty acid called 3-methyl-2-hexenoic acid. (See Figure 4-1.) Sex hormones and androgen sulfates (present only in teens and adults) also contribute to underarm odor.

$$CH_3-CH_2-CH_2-\overset{\overset{CH_3}{|}}{C}=CH-\overset{\overset{O}{\|}}{C}-OH$$

*Figure 4-1: 3-methyl-2-hexenoic acid*

People use deodorants and antiperspirants to control underarm odor. These products work by killing the odor-producing bacteria, slowing bacterial growth, reducing the amount of sweat available to the bacteria, and/or masking the odor with fragrance.

Deodorants contain a perfume to mask body odor and a germicide to kill odor-causing bacteria. The germicide is usually either a long-chain quaternary ammonium salt or a phenol, such as triclosan. (See Figure 4-2.) Antiperspirants contain compounds of aluminum, zinc, and zirconium that constrict the openings of the sweat glands. The common name for the ingredient most often used in antiperspirants is aluminum-zirconium-glycine (AZG).

## Section 3 Background: Control and Conquer

*Figure 4-2: Triclosan*

Excessive sweating in the underarms could signal hyperhidrosis, a medical condition that includes excessive sweating in the underarms, feet, or palms. Another condition, bromidrosis, occurs when excessive sweating in the underarms or feet is accompanied by unpleasant odor. Treatments to control excessive sweating range from prescription antiperspirants to botulinum toxin injections to surgical removal of the sweat glands.

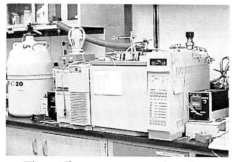

*The sniff port separates and analyzes odor-causing chemicals.*

**Sniffing Out the Problem ▶** Have you ever used a deodorant that didn't fight odor like it was supposed to? If so, you probably stopped using it and bought another brand. Companies developing personal hygiene products try to create products that customers will be happy with and remain loyal to. During product development, researchers at some companies identify the odor-producing chemicals in body fluids, such as sweat, in order to develop technologies that control these unpleasant odors.

Researchers also identify the odors of raw materials used to make the hygiene product so that they can reduce or eliminate any unpleasant odors. Lastly, they test the odor of the final product to make sure its scent will appeal to consumers.

How do scientists find out what odor-producing chemicals are causing a scent? Researchers at The Procter & Gamble Company use the sniff port gas chromatograph/mass spectrometer (GC/MS) system to identify odor-causing chemicals within a substance. The instrument separates and analyzes odorous compounds.

To begin an analysis, the instrument gathers a sample of air from above the test substance. The gas chromatograph separates the air sample into its individual chemical components. Part of the separated sample goes to the mass spectrometer, which identifies the structure of each chemical. Simultaneously, the other part of the separated sample goes to a sniff port, where a scientist sniffs the odor of each chemical component and grades its intensity and character.

# Section 3 Experiment: Life in the Pits

How do deodorants and antiperspirants work? Simulate underarm conditions and find out. In the experiment, yeast is used to simulate bacteria.

## Materials

Per group
- warm water from the teacher
- large Styrofoam® cup
- 250- to 400-mL beaker
- 2, 150-mm test tubes
- wax pencil or labels and permanent marker
- 1.25-mL (¼-teaspoon) and 15-mL (1-tablespoon) measuring spoons
- 1.25 mL (¼ teaspoon) sugar
- 1.25 mL (¼ teaspoon) rapid-rise active dry yeast
- 2 test tube stoppers
- test substance
- 2 large latex balloons
- measuring tape or string and ruler

## Safety and Disposal

According to class policy, empty test tube contents into a solid waste receptacle.

## Procedure

1. Prepare a water bath by filling the Styrofoam cup halfway with warm water from your teacher. Place the cup into the empty beaker to prevent the cup from tipping.

2. Label one test tube "control" and the other one with the name of your test substance. Add 15 mL (1 tablespoon) warm water (from your teacher), 1.25 mL (¼ teaspoon) sugar, and 1.25 mL (¼ teaspoon) yeast to each test tube. Stopper the test tubes and shake for about 20 seconds to mix the contents. Remove the stoppers and place the test tubes in the water bath.

3. Add 1.25 mL (¼ teaspoon) of the test substance to the appropriate test tube, replace the stopper, and shake for about 20 seconds. Remove the stopper and place the test tube back into the water bath.

4. Inflate two balloons about halfway. Use a marker or wax pencil to draw a line around each balloon at its widest point. Deflate, reinflate, and deflate the balloons several times to stretch the latex.

# Section 3 Experiment: Life in the Pits

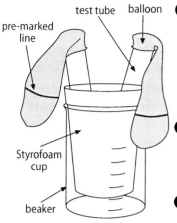

Figure 4-3: Apparatus for yeast test

5. Work with a partner to secure a deflated balloon over the opening of each test tube. Use a measuring tape or a piece of string and ruler to measure the circumference of the balloons at the premarked line. Compare the relative size and quantity of bubbles in the test tubes. Return the test tubes to the water bath. (See Figure 4-3.) Record the appropriate information in the Data Table.

6. Observe the test tubes and balloons, noting the bubbling action. At 15–20 minutes, measure the circumference of the balloon. Fill out the appropriate information in the Data Table.

7. Compare the relative size and quantity of bubbles in the test substance test tube with those in the control test tube, and look at the test tubes prepared by other students who used different test substances.

8. Share your results with the rest of the class. Discuss your observations, noting any differences between groups who used different test substances.

| Data Table for Yeast Test | | |
|---|---|---|
| Test Substance Used: | | |
|  | Balloon Circumference (in centimeters) | Description of Bubbling Action |
| Control (initial reading) | | |
| Control (after ___ minutes) | | |
| Test Substance (initial reading) | | |
| Test Substance (after ___ minutes) | | |

**Question 1:** *How do the experimental conditions model underarm conditions?*

**Question 2:** *Did your test substance affect the growth of yeast? Which test substances used in the class had the most significant effect on the growth of the yeast?*

Chapter 4: When Life Stinks

## Instructor Notes for Section 3

Students simulate underarm bacterial growth by growing yeast. While yeast is a fungus rather than a bacterium, this system serves as a model to explore how ingredients in deodorants and antiperspirants may help to control bacterial growth.

### Time Required
Setup: 5 minutes
Procedure: 35–45 minutes
Cleanup: 10 minutes

### Materials
For Getting Ready
- saucepan or other large container
- warm tap water
- thermometer

For the Procedure
Per group
- warm water
- large Styrofoam cup
- 250- to 400-mL beaker
- 2, 150-mm test tubes
- wax pencil or labels and permanent marker
- 1.25-mL (¼-teaspoon) and 15-mL (1-tablespoon) measuring spoons
- 1.25 mL (¼ teaspoon) sugar
- 1.25 mL (¼ teaspoon) rapid-rise active dry yeast
- 2 test tube stoppers
- 1 of the following test substances:
  - table salt
  - flour
  - antibacterial liquid hand sanitizer
  - liquid soap
  - liquid or solid underarm antiperspirant
  - liquid or solid underarm deodorant
  - vinegar
  - baking soda
  - alum

Instructor Notes for Section 3

- 2 large latex balloons (Use the same size for all groups.)
- measuring tape or string and ruler

## Safety and Disposal

Instruct students to empty test tubes into a solid waste receptacle according to class policy.

## Getting Ready

Break up any solid test substances so students can measure them with a 1.25-mL (¼-teaspoon) measuring spoon.

Just before the experiment, fill a saucepan or large container with warm tap water. Use a thermometer to make sure the water is about 104°F (40°C).

## Procedure Notes

Active dry yeast is a very small fungus in the dormant stage. This experiment uses the yeast to simulate bacteria found in the underarm. Ask students what type of environment will activate the yeast. (Yeast needs moisture, warmth, air, and food.) Review the reaction that occurs in this process. (See Explanation.)

If the data collection portion of the Procedure takes the entire class time, the data analysis can be done on another day.

## Sample Data

| Sample Data Table for Yeast Test | |
|---|---|
| | Balloon Circumference (in centimeters) After 15 Minutes |
| control | 16.4 |
| alum | 5.5 |
| liquid soap | 16.3 |
| antibacterial soap | 14.5 |
| vinegar | 5.0 |

## Answers and Observations

**Question 1:** *How do the experimental conditions model underarm conditions?*

The experimental conditions provide warmth and moisture to mimic underarm conditions.

Chapter 4: When Life Stinks

Instructor Notes for Section 3

**Question 2:** *Did your test substance affect the growth of yeast? Which test substances used in the class had the most significant effect on the growth of the yeast?*

Answers will vary.

## Explanation

When yeast is mixed with warm water and glucose (or other common sugars), the yeast begins to grow. Carbon dioxide ($CO_2$) gas is produced in this process.

$$C_6H_{12}O_6 \xrightarrow{\text{yeast enzyme}} 2\,C_2H_5OH + 2\,CO_2$$
$$\text{glucose} \qquad\qquad \text{ethanol} \quad \text{carbon dioxide}$$

While yeast is used to model bacteria growth in the experiment, note that yeast is a fungus rather than a bacterium. However, both thrive in an environment of moderate temperature, high humidity, and low salt concentration. Instead of producing carbon dioxide, however, underarm bacteria produce malodorous gases as a by-product. Therefore, bacterial growth and resultant underarm odor can be reduced by making the environment less than ideal for bacterial growth. Deodorants contain products that mask body odor and kill odor-causing bacteria. Antiperspirants work by plugging pores in the skin to prevent perspiration from coming to the surface of the skin. Antiperspirants also have a high salt concentration that helps to deter bacterial growth.

# Chapter 5
# Combating and Controlling Body Odor

Chapter 5 Overview .................................................................................................. 100
National Science Education Standards ................................................................... 100
Cross-Curricular Integration .................................................................................... 101

Section 1 Background A: The History of Bathing .................................................... 102
Section 1 Background B: Perfumes and Sensitivity ................................................ 104
Section 1 Experiment: Pick a Fragrance ................................................................. 106
Instructor Notes for Section 1 ................................................................................. 108

Section 2 Background: Scent Masquerade ............................................................. 111
Section 2 Experiment: The All-American Cover-Up ................................................ 113
Instructor Notes for Section 2 ................................................................................. 115

# ● Chapter 5 Overview

**Key Science Topics**
- role of personal hygiene in controlling body odor
- use of deodorants and perfumes to control body odor
- chemistry of deodorants, antiperspirants, and perfumes

As teenagers reach puberty, they become more and more aware of, and concerned about, how other people perceive them. Since the physical changes of puberty include an increase in body odor, covering up body odor and smelling nice to other people can be a challenge to teens.

Chapter 5 explores bathing, fragrances, and deodorants and antiperspirants. Section 1 explores fragrance preferences by asking students to smell and rate several different fragrances and determine gender and class preference patterns. Section 2 provides an overview of how deodorants and antiperspirants work to fight underarm odor and wetness. The Section 2 experiment involves students in exploring the functions of antiperspirants and deodorants.

## National Science Education Standards

This chapter addresses the National Science Education Content Standards for grades 5–8 and 9–12 as described in the following lists.

### Section 1

*Science as Inquiry:*
Abilities Necessary to Do Scientific Inquiry
- Students conduct an investigation and use the evidence gathered to determine whether fragrance preference can be linked to gender.
- Use appropriate tools and techniques to gather, analyze, and interpret data. Students use mathematics to calculate the average rating for each perfume, cologne, and aftershave.

*Science and Technology:*
Abilities of Technological Design
- Students evaluate various perfumes, colognes, and aftershaves to determine whether preferences can be linked to gender.

### Section 2

*Science as Inquiry:*
Abilities Necessary to Do Scientific Inquiry
- Students conduct an investigation into how deodorants work to cover up body odor. Students also design and conduct an investigation into how effective antiperspirants are at controlling moisture.

- Students develop explanations for the different results based on evidence gained from this experiment.

*Science and Technology:*
Abilities of Technological Design
- Students evaluate the effectiveness of antiperspirant in controlling moisture.

## Cross-Curricular Integration

The experiments in this chapter can be integrated with other areas of the curriculum to emphasize the relationships between subjects. Some ideas for cross-curricular integration are listed below.

### Language Arts
- Integrate this experiment with a lesson on using a thesaurus to expand scent vocabulary. Look for adjectives such as musky, fruity, floral, sweet, and fragrant.

### Mathematics
- Tabulate the perfume preferences in Section 1 and study averages, percentages, and ranges.

### Social Sciences
- Study how manufacturers use scent to influence product perception and consumer buying. Compare scents used for products targeted for females and those targeted for males.
- Study the history of the perfume industry and how the crusaders brought perfumes back from the Middle East after their encounters with Arab culture.
- Study other cultures' attitudes on bathing and their use of perfumes and deodorants. Compare the findings with American hygiene regimens.
- Study how fragrances cause allergy symptoms and chemical sensitivities in some people. Learn what can be done to help them.

# ● Section 1 Background A: The History of Bathing

If you enjoy a relaxing bath after a stressful day, you're pretty typical of most people throughout history. Prehistoric people probably waded in rivers or the sea to wash. Ancient Tahitians were so addicted to bathing in their island's streams that they would try to do so several times a day, even if severely ill. Tribal Germans during the time of Julius Caesar were known to bathe in rivers in the middle of winter.

At some point, people began building tubs or other vessels in which they could soak at leisure. A large brick tub or pool, presumably used for ritual bathing, was found in excavations of the city of Mohenjo-Daro in India, which was populated around 2500 BC. One of the oldest true bathtubs is at the royal palace of Knossos in Crete. The palace, which was supposedly built for the legendary King Minos sometime between 1700 and 1400 BC, had one of the most advanced plumbing systems of its day.

We know the ancient Greeks loved bathing because the practice is mentioned often in Homer's *Iliad*. By 400 BC, many Greek towns had public baths where citizens could shower or take a refreshing plunge after working out at the adjoining gymnasium. Although soap is known to have been used in Babylonia as early as 2800 BC, it was probably not used for bathing in ancient times. The 3,500-year-old Ebers papyrus from Egypt mentions a mixture of oils and alkaline salts similar to soap that was used as a skin ointment.

The Romans brought bathing into its golden age. Roman public baths date to around 200 BC, but the time of their greatest splendor was between the first and fourth centuries AD, when elaborate imperial baths, or thermae, were built all over the Empire. Near the end of this period, the city of Rome boasted up to 11 public baths—most of which were built by successive emperors who tried to out-do their predecessors in scale and architecture. The Baths of Caracalla covered nearly 28 acres and could handle over 1,600 bathers at one time. The baths built by the emperor Diocletian may have entertained crowds of up to six thousand. These baths were grand complexes that included theatres, museums, restaurants, and sports facilities. The Romans didn't just go to the baths to get clean—their public baths served as health spas and places for people to socialize and have fun.

A popular misconception holds that people in the Middle Ages did not bathe. In fact, they were not exceptionally dirty and loved to bathe whenever they could. The bedchambers of noblemen often had a half-barrel-shaped tub in which the occupant could soak in hot water, sometimes with perfume or rose petals. Some castles had a room beside the kitchen where the ladies could bathe at parties. Public baths, called stews, were present in many medieval cities for the commoners to use.

For some reason, the popularity of bathing seemed to decline in Europe during the Renaissance and early modern times. By the end of the Middle Ages, large portions of Europe were becoming cleared of forests. Perhaps wood became more scarce and expensive and using it to heat bathwater became less practical. Also, changing religious attitudes of the sixteenth century led many people to frown on public bathing because they thought it encouraged sinful behavior. Some doctors at the time claimed bathing was bad for their patients' health. These habits and notions carried over into the American colonies.

Until the late eighteenth century, routine bathing was not common in America. Bathing at that time would have been a laborious affair. Water had to be heated and the tub filled and emptied with a hand pump or pail. Most colonists washed by taking sponge baths. A wealthy few occasionally went to natural hot springs because they believed the water had curative powers. Gradually, during the first half of the nineteenth century, Americans started to bathe more. At first, mainly the upper classes had tubs and showers in their homes. However, as standards of cleanliness rose and doctors learned more about the spread of disease, middle-class people began to bathe regularly too. This trend was accelerated with the coming of mass production, which lowered the cost of plumbing fixtures, fittings, and valves so that average Americans could afford their own bathrooms.

Today, Americans are serious about cleanliness and very sensitive to body odor. The variety of soaps, deodorants, and luxury bathing items available reflects America's obsession with being clean.

# Section 1 Background B: Perfumes and Sensitivity

Perfumes are commercial products made by blending a variety of different chemical compounds, essential oils, and/or certain animal products. Perfumes do not always smell the same on the skin as they do in the bottle, nor do they smell the same from day to day. Hormonal changes, some medications, temperature, humidity, and even a garlic-laden meal eaten the day before all affect how a perfume smells on a particular person.

A fragrance also changes with time as a person wears it. Perfumes contain three different layers of fragrance, called notes. The top note is made of very small, volatile molecules. When you first apply a fragrance, the nose immediately detects this top note. However, the top note evaporates in about 10 minutes, revealing the middle note, which consists of larger and less volatile molecules. Finally, after up to 45 minutes, the warmth of the skin activates the base note. The base note is made of very large, relatively less volatile molecules. Combined with traces of the first two notes and the chemicals found in the wearer's skin, the base note makes up the distinctive scent of the fragrance, which lasts until it is removed from the skin.

In America, an interpersonal distance of about 18 inches (about 46 cm) is standard between adults. This distance allows personal odors to diffuse so that people usually do not get concentrated whiffs of other people.

Teens often begin using perfumes to prevent body odor from being detected by others entering their interpersonal space. The use of perfume is not limited to females. Colognes and aftershave lotions with distinctively "masculine" scents comprise a considerable portion of the perfume market. Some teens wear a particular perfume or cologne because it makes them feel more self-confident when they socialize with other people. They hope that the scent will make others respond favorably toward them.

Fragrances can play a role in keeping one's mind alert yet relaxed. One research study found that certain fragrances improve a person's concentration on a given task. Another study found that fragrances can help relieve anxiety and stress. Perfumes also play a role in some religious ceremonies. For example, Islamic men and women spray themselves with perfume before each of their five daily prayer sessions.

## Section 1 Background B: Perfumes and Sensitivity

Although most people enjoy wearing perfume or smelling products with fragrance, several studies indicate that 15–30% of the general public have some sensitivity to chemicals, including fragrances. Of these people, more than 80% feel that exposure to fragrances is bothersome. An adverse reaction to fragrance doesn't necessarily result in only a runny nose or sneezing. Many people with fragrance allergies experience dizziness, nausea, fatigue, shortness of breath, nasal and throat soreness, migraines, sinus infections, and asthma.

Some manufacturers offer fragrance-free and unscented versions of their products. However, some of these products contain masking agents that are added to cover up the odors of other chemicals in these products. Unfortunately, some consumers are even sensitive to these masking agents. To address sensitivity to odors in the workplace, some companies have established workplace policies that discourage or ban the use of certain fragrances or perfumes out of respect for the health concerns of their sensitive employees.

*Advertisements for 4711, Cologne's original cologne*

**Cologne from Cologne** ▶ If you've ever shopped for fragrances, you may have seen products called eau de cologne or simply, cologne. As sold today, cologne is a diluted form of perfume—with a 5% concentration of essential oils as compared to perfume's 20–25% concentration.

But the term eau de cologne did not always refer to a diluted perfume. Although the details are sketchy, followers of perfume history believe that a closely-guarded formula for a fragrant, medicinal product was brought to Cologne, Germany, from Italy in the late 1600s or early 1700s. The alcohol-based formula had a 5% concentration of citrus-smelling essential oils such as bergamot, orange flower, and lemon. Marketed as a drinkable cure-all for ailments such as stomachaches and bleeding gums, the formula came to be known by the French name "eau de Cologne," meaning "water of Cologne." Some believed it could even ward off bubonic plague.

Over time, the product was marketed as a fragrance rather than a medicine, and eventually the term eau de cologne became associated with any dilute fragrance.

Chapter 5: Combating and Controlling Body Odor

# ● Section 1 Experiment: Pick a Fragrance

Which types of fragrance do you prefer?

## Materials

Per group
- set of odor strips in test tubes

## Procedure

❶ Each group will be given a set of unknown fragrances in numbered test tubes. Remove a paper strip from a test tube. The fragrance's number appears on the strip. As odor strips are passed around the group, gently wave each strip under your nose. If the odor is not detectable with this method, sniff closer to the strip but don't touch your nose to the strip. To empty your olfactory receptors, sniff your sleeve after you sniff each strip. Trained odor judges call the technique of smelling one's sleeve "going home."

❷ Record your data in the following Data Table. Write down the number found on each strip and a description of the odor, using words like woodsy, musky, sweet, floral, or fruity. Put a check mark in the box representing the rating you give the fragrance.

| Data Table for Rating Fragrance | | | | | | |
|---|---|---|---|---|---|---|
| Odor Strip Number | Description of Odor | 1 Very Pleasing | 2 Pleasing | 3 Average | 4 Not Very Pleasing | 5 Very Displeasing |
| | | | | | | |
| | | | | | | |
| | | | | | | |
| | | | | | | |
| | | | | | | |
| | | | | | | |

Section 1 Experiment: Pick a Fragrance

❸ As a class, compile ratings results in the Class Data Table.

| | Class Data Table | | | | | | | | | | | |
|---|---|---|---|---|---|---|---|---|---|---|---|---|
| | Number of Girls Who Selected the Rating | | | | | | Number of Boys Who Selected the Rating | | | | | |
| Odor Strip Number | 1 | 2 | 3 | 4 | 5 | Average Rating for Girls | 1 | 2 | 3 | 4 | 5 | Average Rating for Boys |
| | | | | | | | | | | | | |
| | | | | | | | | | | | | |
| | | | | | | | | | | | | |
| | | | | | | | | | | | | |
| | | | | | | | | | | | | |
| | | | | | | | | | | | | |
| | | | | | | | | | | | | |

❹ After all students have contributed their data, average the ratings. Calculate the boys' and girls' average ratings for the fragrances.

❺ Analyze the class results from step 4. Look for preference patterns. The teacher will provide information on whether the product is marketed to males or females. How do your class findings relate to the targeted gender for each product? Write a paragraph summarizing your conclusions.

**Question 1:** *Do girls like female perfumes and colognes, while the boys prefer the male aftershaves and colognes? Does the opposite hold true? Or, are you unable to identify any trends in preferences by gender?*

❻ Hold a class discussion to share the results of these analyses. Discuss how companies use fragrance to sell products.

**Question 2:** *Which type of scent (for example, fruity or woodsy) did students, on average, seem to prefer?*

# Instructor Notes for Section 1

Students describe and rate the odors of various perfumes, colognes, and aftershaves. Next, they compile and analyze class results and look for preference trends by gender.

## Time Required
Setup: 15 minutes
Procedure: 30 minutes
Cleanup: 5 minutes

## Materials
For Getting Ready
Per class
- marker
- filter paper or chromatography paper
- scissors or paper cutter
- variety of perfumes, colognes, and aftershaves

 *You can ask students to bring these products from home.*

Per group
- test tubes with stoppers
- beaker or test tube rack to hold the test tubes

For the Procedure
Per group
- set of odor strips in test tubes

## Getting Ready
For each group, place a test tube for each fragrance in a beaker or test tube rack. Use a marker to label each test tube with a number identifying the fragrance. Cut filter paper or chromatography paper into strips that fit inside the test tubes, making enough strips so that each group gets one strip from each test tube. Use a pencil to label each strip with the fragrance's number. Place one or two drops of each fragrance in the test tube or spray the strips with one spray. Stopper the test tubes.

## Procedure Notes
The experiment can be introduced by asking students questions like: Do you have a special fragrance you like to wear? What attracts you to that fragrance?

> Why do you think people wear fragrance—for themselves or for other people? Do people sometimes wear too much fragrance? Point out to students that we're using the industry scent panel's typical odor-detecting method because the odors are produced by known, safe sources at safe concentrations. As discussed in Section 1 of Chapter 2, the wafting technique would be used with potentially dangerous or unknown substances.

## Answers and Observations

**Question 1:** *Do girls like female perfumes and colognes, while the boys prefer the male aftershaves and colognes? Does the opposite hold true? Or, are you unable to identify any trends in preferences by gender?*

Answers will vary.

**Question 2:** *Which type of scent (for example, fruity or woodsy) did students, on average, seem to prefer?*

Answers will vary, but floral and fruity odors seem to have the most universal appeal.

## Explanation

Studies of scent preferences are big business. Perfumes, colognes, and a myriad of personal hygiene products clearly base their success on odor acceptance. Manufacturers are more frequently adding odors to sell items that traditionally do not have odors. Also, unpleasant odors of manufactured items are usually masked by incorporating a stronger, pleasant odor.

Scent is added to clothes detergents and cleaning products to give a "fresh, clean feeling." Even automobile dealers have found that spraying a used car with "odor of new car"—a mixture of oil, leather, and metal scents—will give the potential buyer the subconscious impression of "newness."

A few general rules of odor preference have been noted during extensive preference studies:
- Scents occurring in nature are preferred over synthetic scents.
- People of the same culture generally agree on what smells really bad or really good.
- Some differences in odor preferences are linked to gender or culture.
- Flower and fruit smells generally seem to have a universal appeal.
- Association of a smell with some personal experience may play a part in whether it is liked or not.

*Overhead Master*

## Class Data Table

| Odor Strip Number | Number of Girls Who Selected the Rating | | | | | Average Rating for Girls | Number of Boys Who Selected the Rating | | | | | Average Rating for Boys |
|---|---|---|---|---|---|---|---|---|---|---|---|---|
| | 1 | 2 | 3 | 4 | 5 | | 1 | 2 | 3 | 4 | 5 | |
| | | | | | | | | | | | | |
| | | | | | | | | | | | | |
| | | | | | | | | | | | | |
| | | | | | | | | | | | | |
| | | | | | | | | | | | | |
| | | | | | | | | | | | | |
| | | | | | | | | | | | | |

# Section 2 Background: Scent Masquerade

People use scented deodorants to mask and control body odor, antiperspirants to control body odor and underarm wetness, or deodorant/antiperspirant combination products to achieve both results. Because underarm odor comes from bacterial action on accumulated sweat, daily washing alone often solves many underarm odor problems. However, our culture and the marketing efforts of deodorant/antiperspirant makers help to convince us that these products are necessary.

How do deodorants work? The fragrances used in deodorants are perfumes that can be detected before the underarm odor can be smelled. This process is known as masking. Imagine that a person's nose contains a pegboard with only a hundred holes and that each scent molecule is a peg. When a person smells perfume, the perfume molecules fill up all hundred holes and a signal is sent to the brain, which interprets the signal as perfume. If sweaty underarm molecules are introduced while the pegboard holes are still full, the brain will not get that message until the perfume molecules leave the pegboard. If the perfume molecules keep jamming the smell receptors in the nose, sweat molecules will not be smelled. Some perfumes react only with the smell receptors in the nose that respond to bad odors. This way they block only the bad odors, not the pleasant ones.

Body odor is produced when apocrine sweat is decomposed by bacterial action, so anything that reduces the presence of bacteria also reduces body odor. In addition to masking body odor with perfumes, stick deodorants create a slightly acidic environment on the skin that discourages bacterial growth. Many deodorants also contain antimicrobial agents, such as triclosan, that kill bacteria on contact and offer residual antibacterial effects. Crystal deodorants, or "deodorant stones," which are advertised as alternatives to common stick deodorants, create a salty environment that discourages bacterial growth and acts as a drying agent on the skin.

Deodorant soaps are also available. These soaps contain fragrances that mask body odors. Some deodorant soaps also contain antibacterial agents. However, dermatologists recommend using deodorant soaps only on the "smelly" parts of the body such as the underarm and groin areas. Why? The astringent nature of the soap can be too harsh for parts of the body such as the arms and legs. Teens should consider using moisturizing soaps in these drier areas to keep the skin from becoming dry, itchy, and cracked.

*Tussie Mussie*

**What Is a Tussie Mussie?** ▶ Imagine how smelly your town would be if everyone dumped their garbage into the street. That's what many people in Europe did during the Middle Ages because most cities had no garbage pickup. The odor in these filthy streets was sometimes unbearable. To mask these odors, some men and women held small bundles of fragrant flowers and herbs, called nosegays, under their noses. The flower stems were often wrapped in damp moss to keep them fresh. Another name for a nosegay is tussie mussie or tuzzy muzzy. Tussie means a cluster of flowers and mussie may be a form of the word mossy. In addition to masking odors, people believed that the strong herbs used in tussie mussies protected against airborne germs and the plague.

By the early 1700s, tussie mussies took on a new role in European society with the growing popularity of floriography, the art of sending messages by flowers. Each type of flower had a special meaning. For example, lavender meant devotion, a violet meant faithfulness, and a red rose meant love. Men gave women tussie mussies with specific flowers to communicate individualized messages of courtship and love.

Today, the popularity of tussie mussies is on the rise with the revival of Victorian romanticism. The term tussie mussie can refer to both the floral bouquets and to the containers designed to hold them. Tussie mussie containers are small cone- or cup-shaped holders made from a variety of materials and are sold to hold flowers for proms and weddings.

Another strategy for controlling body odor is to use antiperspirants to reduce the amount of perspiration available to the bacteria. The main ingredients in antiperspirants are aluminum and zinc salts and alcohol, with the most common active ingredient being aluminum-zirconium-glycine (AZG).

Antiperspirants constrict the openings of the sweat glands to retard the flow of perspiration to the surface of the skin for as long as several weeks. Although the exact mechanism of this action is unclear, the highly charged aluminum ions may bind with natural compounds in the sweat glands and retard the flow of sweat. Additionally, the high concentration of aluminum or zinc salts together with the alcohol in antiperspirants provide an astringent action, causing the pores themselves to shrink and reducing the amount of sweat released. Finally, solid antiperspirants typically contain powders, such as baking soda and talc, to absorb moisture. Antiperspirants also have a slight antiseptic action as a result of various chemicals added for this purpose. Because antiperspirants actually modify the body's sweating function, they are technically considered a drug.

In 1902, the first antiperspirant was introduced under the trade name Everdry. Being primarily a solution of aluminum chloride, it was very effective, but it was also very messy, irritated the skin, and often destroyed the underarm area of the wearer's clothing! Today aluminum chlorohydrate or other aluminum salts are often used. Since most commercial deodorants are also antiperspirants, American teens often choose a combination deodorant/antiperspirant to mask odor and reduce perspiration.

# Section 2 Experiment: The All-American Cover-Up

Like most American teens, you probably use an antiperspirant or deodorant, but do you know how it works?

## Materials

Part A
Per group
- set of 5 odor cards

Part B
Per student
- Fortune Teller Fish
- antiperspirant (brought from home)
- other materials needed for your experiment (brought from home)

## Procedure
### Part A: Masking Odor

❶ Gently wave each odor card under your nose. Describe the odors and your predictions of their identities in the Data Table.

| | Data Table for Masking Odor | |
|---|---|---|
| | Description of Odor | Likely Identity of Odor |
| Card 1 | | |
| Card 2 | | |
| Card 3 | | |
| Card 4 | | |
| Card 5 | | |

**Question 1:** *One of the cards has been treated only with liquid smoke, and the others have been treated with liquid smoke and different masking odors. Which card has been treated only with liquid smoke?*

❷ Share and discuss observations with the rest of the class.

Chapter 5: Combating and Controlling Body Odor

## Section 2 Experiment: The All-American Cover-Up

**Question 2:** *Did the masking odors tend to draw your attention away from the smoke odor? Discuss whether or not some fragrances worked better than others at covering the odor.*

### Part B: Design Your Own Experiment

1. Design and record an experiment that will test whether antiperspirants work to control wetness. Use a Fortune Teller Fish as the moisture indicator in your experiment. (You may wish to review the Section 2 experiment in Chapter 3.)

2. Have your experiment okayed by your teacher. Bring the materials you need from home and try your experiment. Record your results.

3. Compare your results with those of the rest of the class.

**Question 3:** *Did you have a control in your experiment? What were the variables?*

**Question 4:** *How was your experiment similar to those designed by the rest of the class and how was it different?*

**Question 5:** *Were your results the same or different than the results obtained by most of the class? Why?*

**Question 6:** *Would you use the results of your test as a basis for buying a particular antiperspirant? Why or why not?*

# Instructor Notes for Section 2

In Part A of this experiment, students evaluate how fragrances cover smoke odor and learn how deodorants work in masking underarm odor. Part B asks students to use the Fortune Teller Fish from Chapter 2 to design an experiment assessing the effectiveness of an antiperspirant of their choice.

## Time Required

Setup: 15 minutes
Procedure: Part A, 30 minutes
    Part B, time will vary
Clean up: 5 minutes

## Materials

For Getting Ready
- 5 index cards for each group of students
- cotton swabs
- liquid smoke
- four different brands of perfume, cologne, or scented deodorant
- (optional) zipper-type plastic bags

For Part A
Per group
- set of 5 odor cards

For Part B
Per student
- Fortune Teller Fish

 *Fortune Teller Fish (PR9906S for 30 or PR9906L for 144) can be purchased from Terrific Science Books, Kits, & More; 866/438-6724; http://www.tsbkm.com.*
- antiperspirant (brought from home by student)
- other materials needed for the experiment (brought from home by student)

## Getting Ready

Prepare the odor cards no more than 1–2 hours before the experiment since the odors will fade with time. For each group of students, label a set of odor cards 1–5. Dip a cotton swab into liquid smoke and use the swab to draw a dime-sized circle in the center of each card. Next to the liquid smoke circle on

**Chapter 5: Combating and Controlling Body Odor**

> **Instructor Notes for Section 2**

cards 2–5, add a dime-sized circle of a different brand of perfume, cologne, or scented deodorant. Keep a key to the card number and the identity of each odor used.

To prevent contamination, do not stack the scented index cards. You can experiment with preparing the cards the day before the experiment and storing them in separate zipper-type plastic bags; how much odor remains on the cards will depend on the particular substances you are using. Alternatively, you can allow students to prepare their own cards for this experiment or have each group prepare another group's cards.

## Procedure Notes

Introduce the experiment with the following discussion: Why are most deodorants scented? Why do people often use an air freshener spray after cooking fish or in bathrooms? (The more pleasant odors cover up, or mask, the unpleasant odors.)

## Answers and Observations

**Question 1:** *One of the cards has been treated only with liquid smoke, and the others have been treated with liquid smoke and different masking odors. Which card has been treated only with liquid smoke?*

Card 1, the control, was treated only with liquid smoke. The other cards were treated with liquid smoke as well as different fragrances.

**Question 2:** *Did the masking odors tend to draw your attention away from the smoke odor? Discuss whether or not some fragrances worked better than others at covering the odor.*

Answers will vary depending on the fragrances used. Some fragrances will probably work better than others in masking the smoke odor.

**Question 3:** *Did you have a control in your experiment? What were the variables?*

Answers will vary.

**Question 4:** *How was your experiment similar to those designed by the rest of the class and how was it different?*

Answers will vary depending on the individual experimental designs. One possible experiment would be to first observe the behavior of the Fortune Teller Fish on a hand. The fish should curl because it absorbs the moisture on the hand. Next, apply antiperspirant to the hand and allow it to dry. Place the Fortune Teller Fish on the hand again and observe its behavior. The fish should not curl as much because the antiperspirant prevents moisture.

**Question 5:** *Were your results the same or different than the results obtained by most of the class? Why?*

Answers will vary. In general, antiperspirants control wetness, and their effectiveness probably doesn't differ much from brand to brand. However, individual student results may be different for many reasons. For example, hand moisture amounts vary from person to person. Also, in some cases, the antiperspirant may not have been applied thoroughly enough to be a true test of its effectiveness.

**Question 6:** *Would you use the results of your test as a basis for buying a particular antiperspirant? Why or why not?*

Answers will vary.

## Explanation

Part A of the experiment demonstrates that fragrances can be used to mask unpleasant odors just like scented deodorants are used to mask body odor. The fragrance in these deodorants introduces pleasant-smelling molecules to the nose. These molecules occupy the odor receptors in our nose so that unpleasant body odors will not be perceived.

Students design their own experiments with Fortune Teller Fish in Part B to investigate the effectiveness of antiperspirants in controlling moisture. Fortune Teller Fish are made of a special type of cellophane that absorbs moisture. As a result, the fish curls when moisture is present. Antiperspirant placed on the hand should reduce the amount of moisture the fish comes in contact with, thereby minimizing the curling.

# Bibliography

ABC News Website. Sci/Tech. http://abcnews.go.com (accessed October 2002).

Ackerman, D. *A Natural History of the Senses;* Random House: New York, 1990.

Affinity Laboratory Technologies, Inc., Website. Halitox Background. http://www.altcorp.com/AffinityLaboratory/halitoxbackground.htm (accessed November 2002).

Agency of Industrial Science and Technology Website. Public Relations Magazine; 2002 No. 4 Spring; Molecular Basis of Odor Discrimination in Olfaction. http://www.aist.go.jp (accessed October 2002).

Alternative Fuels Data Center Website. Alternative Fuels; Natural Gas General Information. http://www.afdc.doe.gov (accessed October 2002).

Altruis Biomedical Network Website. History. http://www.oral-products.com (accessed October 2002).

Alvin, V.; Silverstein, R. *Smell, The Subtle Sense;* Morrow Junior: New York, 1992.

Amoore, J.E. *The Molecular Basis of Odor;* Charles C. Thomas: Springfield, IL, 1970.

AromaWeb Website. History of Aromatherapy. http://www.aromaweb.com (accessed October 2002).

AromaWeb Website. What Is Aromatherapy? http://www.aromaweb.com (accessed October 2002).

AromaWeb Website. What Are Essential Oils? http://www.aromaweb.com (accessed October 2002).

Bartleby.com Website. Columbia Encyclopedia; Eau de Cologne. http://www.bartleby.com (accessed October 2002).

Baxter, R. Mouthwash: What's In It for You. *ChemMatters.* December 1996, 6–8.

Bishop, M. *The Middle Ages;* American Heritage: New York, 1970.

Bloomfield, M. *Chemistry and the Living Organism,* 5th ed.; Wiley & Sons: New York, 1992.

Brooks, G. Behind the Veil. *Scholastic Update.* October 22, 1993, 12–14.

Bushman, R.L.; Bushman, C.L. The Early History of Cleanliness in America. *The Journal of American History.* March 1988, *74*(4), 1213–1238.

Celestial Touch Website. Essential Oils; Neroli. http://www.celestialtouch.com (accessed October 2002).

Clifford, C. Home Remedies Even Doctors Use. *McCall's,* August 1992, 48.

Cologne Website. 4711 Museum (requires translation). http://www.cologneweb.com (accessed October 2002).

Colonial Williamsburg Website. Taking the Cure: Colonial Spas, Springs, Baths, and Fountains of Health. http://www.history.org (accessed October 2002).

Colonial Williamsburg Website. To Bathe or Not to Bathe: Coming Clean in Colonial America. http://www.history.org (accessed October 2002).

Columbia Gas Website. Safety Information; Mercaptan. http://www.columbiagaspamd.com (accessed October 2002).

Corbin, A. *The Foul and the Fragrant;* Harvard University: Cambridge, MA, 1986.

Cox Newspapers Website. Washington Bureau; Shelley Emling; Pentagon Seeking the Ultimate Stink Bomb (01-10-02). http://www.coxnews.com (accessed October 2002).

*CRC Handbook of Chemistry and Physics,* 81st ed.; Lide, D.R., Ed.; CRC Press: Boca Raton, FL, 2000.

Cyber-Bohemia Website. Mediterranean Baths; Early Greek and Roman Baths. http://www.cyberbohemia.com (accessed October 2002).

Cyber-Bohemia Website. Mediterranean Baths; Mass Bathing: The Roman Balnea and Thermae. http://www.cyberbohemia.com (accessed October 2002).

DaeDalus Informatics Website. The Emergence of the Minoan Civilization; Bathroom in the Queen's Apartment. http://www.daedalus.gr/DAEI/THEME/Minos20.html (accessed October 2002).

D.D. Williamson Website. Caramel Color in Carbonated Soft Drinks. http://ddwilliamson.com (accessed October 2002).

Dentist.net Website. Treat Bad Breath; Worried About Bad Breath? http://www.dentist.net (accessed October 2002).

Desert USA Website. Kangaroo Rats. http://www.desertusa.com (accessed October 2002).

Drugstore.com Website. Binaca. http://www.drugstore.com (accessed October 2002).

Drugstore.com Website. Breath Remedy Non-Aerosol Tongue Spray. http://www.drugstore.com (accessed October 2002).

East Carolina University Website. Parosmia and Phantosmia. http://personal.ecu.edu/wuenschk/parosmia.htm (accessed October 2002).

EBSCO Host Website. Funk & Wagnall's New World Encyclopedia; Baths. http://www.ebscohost.com (accessed October 2002).

Edersbow Website. OnLine Magazine; May 1999; Can We Eliminate Human Scent? http://www.edersbow.com (accessed October 2002).

eMedicine Website. Disorders of Taste and Smell. http://www.emedicine.com (accessed October 2002).

EMuseum (Minnesota State University) Website. Archeology; Sites; Middle East; Mahenjo-Daro. http://emuseum.mnsu.edu (accessed October 2002).

Engen, T. *Odor Sensation and Memory;* Praeger: New York, 1991.

Engen, T. *The Perception of Odor;* Academic Press (Harcourt Brace Jovanovich): New York, 1982.

Environmental Molecular Sciences Laboratory Website. Medical Technology Brief: Electronic/Artificial Noses. http://www.emsl.pnl.gov:2080/ (accessed October 2002).

Erb, R.C. *The Common Scents of Smell;* World: Cleveland, OH, 1968.

The Essential Oil Company Website. Absolutes and Exotic Oils. http://www.essentialoil.com (accessed October 2002).

Family Education Network Website. Encyclopedia; Sweat. http://www.infoplease.com (accessed October 2002).

Fisher, B.E. Scents & Sensitivity. *Environmental Health Perspectives.* December 1998, *106*(12), A594–A599.

Fragranced Products Information Network Website. The Fragrance Industry. http://www.ameliaww.com/fpin/fpin.htm (accessed October 2002).

Fragranced Products Information Network Website. Heath Conditions Effected by Fragrances. http://www.ameliaww.com/fpin/fpin.htm (accessed October 2002).

*Fun with Chemistry: A Guidebook of K–12 Activities;* Sarquis, A.M., Sarquis, J.L., Eds.; Institute for Chemical Education: Madison, WI, 1991; Vol. 1.

*Fun with Chemistry: A Guidebook of K–12 Activities;* Sarquis, A.M., Sarquis, J.L., Eds.; Institute for Chemical Education: Madison, WI, 1993; Vol. 2.

Furtado, T.; Muniz, B. There's No Such Thing as Smelly Feet. *Health.* May/June 1993, *7*(3), 36–37.

Guardian Unlimited Website. Police Sniff Out Mother of All Stink Bombs. http://www.observer.co.uk (accessed October 2002).

Guenther, E. *The Essential Oils;* Van Nostrand: New York, 1949; Vol. 3.

Guenther, E. *The Essential Oils;* Van Nostrand: New York, 1952; Vol. 5.

Guyton, A.C. *Textbook of Medical Physiology,* 7th ed.; W.B. Saunders: Philadelphia, 1986.

Health Central Website. General Health Encyclopedia; Sweating, excessive. http://www.healthcentral.com/mhc/top/003218.cfm (accessed October 2002).

Hill, J.W.; Kolb, D.K. *Chemistry for Changing Times,* 7th ed.; Prentice-Hall: Englewood Cliffs, NJ, 1995.

Home Solutions News Website. Wizard. http://www.homesolutionsnews.com (accessed October 2002).

Homework UK Website. Nature; Elephant. http://www.homeworkuk.fsnet.co.uk (accessed October 2002).

How Stuff Works Website. How Sweat Works. http://www.howstuffworks.com (accessed October 2002).

How Stuff Works Website. What Is Activated Charcoal and Why Is it Used in Filters? http://www.howstuffworks.com (accessed October 2002).

Howard Hughes Medical Institute Website. Seeing, Hearing, and Smelling the World; The Mystery of Smell. http://www.hhmi.org (accessed October 2002).

Howard, P.J. *Owner's Manual for the Brain;* Bard: Austin, TX, 1994.

Hunt, R. The Benefits of Scent Evidence. *FBI Law Enforcement Bulletin.* November 1999, *68*(11), 15–18.

Instituto de Fisiologìa Celular Website. A Brief Introduction to the Brain; The Neuron. http://ifcsun1.ifisiol.unam.mx/Brain/neuron.htm (accessed October 2002).

The Internet Classics Archive Website. Homer; The Iliad. http://classics.mit.edu (accessed October 2002).

Interscan Corporation's Halimeter® Website. Product Information; Measure Bad Breath Scientifically. http://www.halimeter.com (accessed November 2002).

Jolique Website. Browse the Archives; The Origins of Eau de Cologne. http://www.jolique.com (accessed October 2002).

Kauffman, G.B. Chemical Aspects of Antiperspirants and Deodorants. *Journal of Chemical Education.* 1993, *70*(3), 242–244.

KET's Distance Learning Website. History of Ancient Roman Baths; Page 1. http://dl.ket.org (accessed October 2002).

Kidzworld Media Website. NASA Sniffer. http://www.kidzworld.com (accessed October 2002).

Kimball's Biology Pages Website. Olfaction; The Sense of Smell. http://users.rcn.com/jkimball.ma.ultranet/BiologyPages (accessed October 2002).

King, J. *Scents: Chic Simple Components;* Alfred A. Knopf: New York, 1993.

Lake Forest College Website. Minoan Crete. http://www.lfc.edu (accessed October 2002).

Leffingwell & Associates Website. Olfaction—A Review; Olfaction—Page 4. http://www.leffingwell.com (accessed October 2002).

Lewis, R. The Bugs Within Us. *FDA Consumer.* 1992, *4*, 37–42.

Lucile Packard Children's Hospital Website. Foreign Bodies in Ear, Nose, and Airway. http://www.lpch.org (accessed November 2002).

Macalester College Website. Nasal; The Olfactory System: Anatomy and Physiology. http://www.macalester.edu (accessed October 2002).

Marieb, E.N. *Human Anatomy and Physiology;* Benjamin/Cummings: Redwood City, CA, 1989.

Mark H. Habermeyer D.D.S. Website. History of Dentistry. http://www.dentistry4u.com (accessed October 2002).

Martini, F.H. *Fundamentals of Anatomy and Physiology;* Prentice-Hall: Upper Saddle River, NJ, 1998.

McGee, H. *On Food and Cooking, The Science and Lore of the Kitchen;* Collier: New York, 1984.

*The Merck Index,* 12th ed.; Budavari, S., Ed.; Merck & Co.: Whitehouse Station, NJ, 1996.

*Medical Aspects of Harsh Environments;* Pandolf, K.; Burr, R.; Wenger, C.; Pozos, R., Eds.; Borden Institute: Washington DC, 2001; Vol. 1.

Mesmer, K. Science Parties. *Science Scope.* 1995, 44–45.

Moncrieff, R.W. *Odor Preferences;* John Wiley: New York, 1966.

Muir, J. *Bizarre and Beautiful Noses;* John Muir: Sante Fe, NM, 1993.

National Association of Biology Teachers Website. Olfactory Fatigue and Memory. www.nabt.org/sup/publications/nlca/nlca.htm (accessed October 2002).

National Geographic Website. U.S. Military Is Seeking Ultimate "Stink Bomb." http://news.nationalgeographic.com (accessed October 2002).

Newswise Website. MedNews. Allergies: Culprit Could Be in Cosmetic Bag (American Academy of Dermatology; March 11, 2000). http://www.newswise.com (accessed October 2002).

Nutraceutic Website. Aromatherapy. http://www.nutraceutic.com (accessed October 2002).

Odor-Eaters Website. Rotten Sneaker Contest. http://www.odor-eaters.com (accessed October 2002).

Ohio Valley Search and Rescue, Inc., Website. Vocabulary. http://www.vsar.org (accessed November 2002).

Oliver, D.L. *Ancient Tahitian Society, Volume 1 Ethnography;* University Press of Hawaii: Honolulu, 1974.

On the Scent of a Better Day at Work. *New Scientist.* March 2, 1991, *129*(1758), 18.

Oxford English Dictionary Website. http://dictionary.oed.com (accessed October 2002).

Polsdorfer, J.R. Anosmia. In *Gale Encyclopedia of Medicine;* 2nd. ed.; Longe, J.L., Ed.; Gale Group: Detroit MI, 2002; Vol. 1.

Ramsey, M.L. Foot Odor: How to Clear the Air. *The Physician and Sportsmedicine.* August 1996, *24*(8).

Riverburg, R. Sweat: This Gland Is Your Gland. *Star Tribune* (Minneapolis, MN), September 4, 1994.

Rosin, D. *The Sinus Sourcebook;* Lowell House: Los Angeles, 1998.

Rosenberg, M. The Science of Bad Breath. *Scientific American.* April 2002, *286*(4), 72–79.

SaskEnergy Incorporated Website. Safety; What Is Natural Gas? http://www.saskenergy.com (accessed October 2002).

Sarquis, A.M.; Sarquis, J.L.; Williams, J.P. "Fortune-Telling Fish," *Teaching Chemistry with TOYS;* McGraw-Hill: New York, 1994.

Sarquis, A.M. "Identifying Substances by Smell," *Exploring Matter with TOYS: Using and Understanding the Senses;* McGraw-Hill: New York, 1997.

Sarquis, J.L.; Hogue, L.M.; Sarquis, A.M.; Woodward, L. "A Cool Phase Change," *Investigating Solids, Liquids, and Gases with TOYS: States of Matter and Changes of State;* McGraw-Hill: New York, 1997.

Schlosberg, S. A Breath of Fresh Air. *Health.* March 1998, *12*(2), 42–44.

Seeley, R.R.; Stephens, T.D.; Tate, P. *Anatomy and Physiology;* C.V. Mosby: St. Louis, 1989.

Selinger, B. *Chemistry in the Marketplace,* 4th ed.; Harcourt Brace Jovanovich: Marrickville, Australia, 1989.

The Semiotic Review of Books Website. Archives; Harkin, M.; Things the Nose Knows. http://www.chass.utoronto.ca (accessed October 2002).

Simple Soap Website. History of Soapmaking. http://www.simplesoap.com (accessed October 2002).

Simple Solutions Corporation Website. Cooking Dictionary; Limburger Cheese. http://www.simpleinternet.com (accessed October 2002).

The Soap and Detergent Association Website. Cleaning Products Overview; History. http://www.sdahq.org (accessed October 2002).

The Soap and Detergent Association Website. Health and Safety; Cleaning for Health. http://www.sdahq.org (accessed October 2002).

Social Issues Research Centre Website. Publications; The Smell Report. http://www.sirc.org (accessed October 2002).

The Society of Chiropodists and Podiatrists Website. Common Foot Problems; Sweaty Feet. http://feetforlife.org (accessed October 2002).

The Society of Thoracic Surgeons Website. Patient Information; Hyperhidrosis. http://www.sts.org (accessed October 2002).

Steen, E.B.; Montagu, A. *Anatomy and Physiology,* Vol. 1; Barnes & Noble: New York, 1959.

Stuller, J. Cleanliness Has Only Recently Become a Virtue. *Smithsonian.* February 1991, 126.

The Sunshine Project Website. Publications; Non-Lethal Weapons Research in the U.S.: Calmatives and Malodorants (Backgrounder #8). http://www.sunshine-project.org (accessed October 2002).

Symmetry Website. Catalog; Herbal; Spearmint Breath Spray. http://www.go-symmetry.com (accessed October 2002).

Takagi, S. *Human Olfaction;* University of Tokyo: Tokyo, 1989.

*Tasting and Smelling: Handbook of Perception Series,* Vol. 6A; Carterette, E.C., Friedman, M.P., Eds.; Academic Press (Harcourt Brace Jovanovich): New York, 1978.

Tel Aviv University Website. Bad Breath Research. http://www.tau.ac.il/~melros/ (accessed November 2002).

Tennessee Deer Hunting Website. Huntën Articles; Making Sense of Scent. http://www.tndeer.com (accessed October 2002).

Teter, H.E.; Bain, T.D.; Fitzpatrick, F.L. *Living Things;* Holt, Rinehart, and Winston: New York, 1985.

Theplumber.com Website. History of Plumbing in America. http://www.theplumber.com (accessed October 2002).

Thorndike, L. Sanitation, Baths, and Street-Cleaning in the Middle Ages and Renaissance. *Spectrum.* April 1928, *3*(2), 192–203.

Transportation Safety Institute Website. TSI Divisions; Pipeline Safety Division; Publications; Guidance Manual for Operators of Small Natural Gas Systems; Chapter IV: Leak Detection. http://www.tsi.dot.gov (accessed October 2002).

Ultimate Wedding Website. Resources; Flowers; What the Heck Is a Tussy Mussy? http://www.ultimatewedding.com (accessed October 2002).

United States Police Canine Association Website. Training; Scent as Forensic Evidence and its Relationship to the Law Enforcement Canine. http://www.uspcak9.com (accessed November 2002).

United States Police Canine Association Website. Articles; Training; Scent—K9's Reason for Being. http://www.uspcak9.com (accessed November 2002).

United States Search and Rescue Task Force Website. Dogs in Search & Rescue. http://www.ussartf.org (accessed November 2002).

University of Alberta Website. Harappan Culture (Lecture of C.S. Mackay). http://www.ualberta.ca (accessed October 2002).

University of California, San Diego Medical Center, Department of Surgery Website. ENT/Otolaryngology; Terence M. Davidson, M.D., F.A.C.S.; Ambulatory Healthcare Pathways for Ears, Nose, and Throat Disorders. http://www-surgery.ucsd.edu (accessed October 2002).

University of California, San Diego Medical Center, Department of Surgery Website. ENT/Otolaryngology; Terence M. Davidson, M.D., F.A.C.S.; Nasal Dysfunction Clinic. http://www-surgery.ucsd.edu (accessed October 2002).

University of California, San Diego Website. Nasal Dysfunction Clinic; Links to Other Sites About Smell. http://www.ucsd.edu (accessed October 2002).

University of Pennsylvania Website. Pennsylvania Current (February 12, 1998); Electronic Nose Sniffs Out Infections. http://www.upenn.edu/pennnews/current/archives.html (accessed October 2002).

University of Washington Website. Our Chemical Senses: Olfaction. http://faculty.washington.edu/chudler/chems.html (accessed October 2002).

University of Wisconsin—Madison Chemistry Professor Bassam Z. Shakhashiri Website. Chemical of the Week; Chem 103; Methane. http://scifun.chem.wisc.edu (accessed October 2002).

Victorian Bride Website. Tussie Mussie; Truly Victorian Tussies. http://www.victorianbride.com (accessed October 2002).

Vigilante Ventures Website. Bathtubs. http://vigilanteventures.com/trivia/bathtub.htm (accessed October 2002).

Wayne State University, College of Science Website. Psychology Department; HyperText Psychology—Senses/Smell. http://sun.science.wayne.edu (accessed October 2002).

White, L., Jr. The Historical Roots of Our Ecological Crisis. *Science.* March 10, 1967, *155*, 1203–1207.

Whitfield, P.; Stoddart, D.M. *Hearing, Taste, and Smell;* Torstar: New York, 1984.

Williams, M. Dark Ages and Dark Areas: Global Deforestation in the Deep Past. *Journal of Historical Geography.* 2000, *26*(1), 28–46.

Winter, R. *The Smell Book;* J.B. Lippincott: Philadelphia, 1976.

Wired News Website. News; Robo Lobster to Sniff Out Mines. http://www.wired.com (accessed October 2002).

Young Living Essential Oils Website. Crash Course; Ancient History and Modern Rediscovery. http://www.therealessentials.com (accessed October 2002).